REA's Test Prep Books Are The Best!

(a sample of the hundreds of letters REA receives each year)

“ I studied this guide exclusively and passed the [CLEP Introductory Sociology] test with 12 points to spare. ”

Student, Dallas, TX

“ This [REA] book was much better than reading a college textbook. Get it, it's worth it! By the way, I passed [the CLEP test] with flying colors!!! ”

Student, Poughkeepsie, NY

“ Your book was such a better value and was so much more complete than anything your competition has produced — and I have them all! ”

Teacher, Virginia Beach, VA

“ Compared to the other books that my fellow students had, your book was the most useful in helping me get a great score. ”

Student, North Hollywood, CA

“ Your book was responsible for my success on the exam, which helped me get into the college of my choice... I will look for REA the next time I need help. ”

Student, Chesterfield, MO

“ Just a short note to say thanks for the great support your book gave me in helping me pass the test... I'm on my way to a B.S. degree because of you! ”

Student, Orlando, FL

(more on next page)

(continued from front page)

" I just wanted to thank you for helping me get a great score on the AP U.S. History exam... Thank you for making great test preps! "

Student, Los Angeles, CA

" Your *Fundamentals of Engineering Exam* book was the absolute best preparation I could have had for the exam, and it is one of the major reasons I did so well and passed the FE on my first try. "

Student, Sweetwater, TN

" I used your book to prepare for the test and found that the advice and the sample tests were highly relevant... Without using any other material, I earned very high scores and will be going to the graduate school of my choice. "

Student, New Orleans, LA

" What I found in your book was a wealth of information sufficient to shore up my basic skills in math and verbal... The section on analytical ability was excellent. The practice tests were challenging and the answer explanations most helpful. It certainly is the *Best Test Prep for the GRE!* "

Student, Pullman, WA

" I really appreciate the help from your excellent book. Please keep up the great work. "

Student, Albuquerque, NM

" I am writing to thank you for your test preparation... your book helped me immeasurably and I have nothing but praise for your *GRE* preparation. "

Student, Benton Harbor, MI

THE BEST TEST PREPARATION FOR THE

CLEP

Precalculus

Betty Travis, Ph.D.
Professor of Mathematics
University of Texas at San Antonio
San Antonio, Texas

Research & Education Association
Visit our website at
www.rea.com

Research & Education Association
61 Ethel Road West
Piscataway, New Jersey 08854
E-mail: info@rea.com

The Best Test Preparation for the
CLEP PRECALCULUS EXAM

Printed in the United States of America

Library of Congress Control Number 2008932122

ISBN-13: 978-0-7386-0174-8
ISBN-10: 0-7386-0174-8

H-08

About Our Author

Dr. Betty Travis is a Professor in the Department of Mathematics at the University of Texas at San Antonio (UTSA). She has served as Associate Dean of Academic Affairs and Interim Associate Dean for Graduate Studies and Administration in the College of Sciences and Engineering at UTSA and Dean of Academic Affairs at Santa Barbara City College. Dr. Travis graduated Magna Cum Laude with a B.A. in Mathematics from St. Mary's University. She also received her M.S. in Mathematics from St. Mary's University and her Ph.D. in Mathematics Education from The University of Texas at Austin.

Dr. Travis was named to "Who's Who Among America's Teachers," nominated for the San Antonio Women's Hall of Fame, and named one of "2000 Notable American Women." She was also recognized as a "Million Dollar Scholar" by UTSA and nominated for the Louise Hay Award for Outstanding Achievement by a Woman in Mathematics Education. She is the winner of the President's Distinguished Achievement Award in Recognition of Excellence in University Service by UTSA and the AMOCO Teaching Award at UTSA. Dr. Travis has served on numerous state and national committees and is a former President of the Texas Council of Teachers of Mathematics and the Alamo District Council of Teachers of Mathematics.

Dr. Travis has extensive experience in grant management, having written and received approximately $4.5 million in grants from the National Science Foundation, the Department of Education, the Office of Naval Research, the Calculus Consortium for Higher Education and the Texas Higher Education Coordinating Board.

About Research & Education Association

Founded in 1959, Research & Education Association (REA) is dedicated to publishing the finest and most effective educational materials—including software, study guides, and test preps—for students in middle school, high school, college, graduate school, and beyond.

Today REA's wide-ranging catalog is a leading resource for teachers, students, and professionals.

We invite you to visit us at *www.rea.com* to find out how "REA is making the world smarter."

Acknowledgements

In addition to our author, we would like to thank Larry B. Kling, Vice President, Editorial, for his overall guidance, which brought this publication to completion; Pam Weston, Vice President, Publishing, for setting the quality standards for production integrity and managing the publication to completion; Diane Goldschmidt, Senior Editor, for editorial project management; Mel Friedman, Lead Math Editor, for mathematics review; Alice Leonard, Senior Editor, for preflight editorial review; Jeff LoBalbo, Senior Graphic Designer, for coordinating pre-press electronic file mapping; Christine Saul, Senior Graphic Designer, for designing our cover.

We also extend special thanks to Sandra Rush for copyediting, Ellen Gong for proofreading, and Aquent Publishing Services for typesetting this edition.

CONTENTS

CLEP PRECALCULUS
Study Schedule

The following study schedule allows for thorough preparation for the CLEP Precalculus exam. Although it is designed for four weeks, it can be reduced to a two-week course by collapsing each two-week period into one. Be sure to set aside enough time—at least two hours each day—to study. No matter which study schedule works best for you, the more time you spend studying, the more prepared and relaxed you will feel on the day of the exam.

Week	Activity
1	Read and study Chapter 1 of this book, which will introduce you to the CLEP Precalculus exam. Then take Practice Test 1 to determine your strengths and weaknesses. Assess your results by using our raw score conversion table. You can then determine the areas in which you need to strengthen your skills.
2 & 3	Carefully read and study the CLEP Precalculus review material included in Chapters 2 through 7 in this book.
4	Take Practice Test 2 and carefully review the explanations for all incorrect answers. If there are any types of questions or particular subjects that seem difficult to you, review those subjects by again studying the appropriate sections of the CLEP Precalculus review chapters.

Note: If you care to, and time allows, retake Practice Tests 1 and 2. This will help strengthen the areas in which your performance may still be lagging and build your overall confidence.

CHAPTER 1
Passing the CLEP Precalculus Exam

Chapter 1
Passing the CLEP Precalculus Exam

ABOUT THIS BOOK

This book provides you with complete preparation for the CLEP Precalculus exam. Inside you will find a targeted review of the subject matter, as well as tips and strategies for test taking. We also give you two practice tests, featuring content and formatting based on the official CLEP Precalculus exam. Our practice tests contain every type of question that you can expect to encounter on the actual exam. Following each practice test you will find an answer key with detailed explanations designed to help you more completely understand the test material.

All CLEP exams are computer-based. As you can see, the practice tests in our book are presented as paper-and-pencil exams. The content and format of the actual CLEP subject exams are faithfully mirrored. Later in this chapter you'll find a detailed outline of the format and content of the CLEP Precalculus exam.

ABOUT THE EXAM

Who takes CLEP exams and what are they used for?

CLEP (College-Level Examination Program) examinations are typically taken by people who have acquired knowledge outside the classroom and wish to bypass certain college courses and earn college credit. The CLEP is designed to reward students for learning—no matter where or how that knowledge was acquired. The CLEP is the most widely accepted credit-by-examination program in the country, with more than 2,900 colleges and universities granting credit for satisfactory scores on CLEP exams.

Although most CLEP examinees are adults returning to college, many graduating high school seniors, enrolled college students, military personnel, and international students also take the exams to earn college credit or to demonstrate their ability to perform at the college level. There are no prerequisites, such as age or educational status, for taking CLEP examinations. However, because policies on granting credits vary among colleges, you should contact the particular institution from which you wish to receive CLEP credit.

There are two categories of CLEP examinations:

1. **CLEP General Examinations,** which are five separate tests that cover material usually taken as requirements during the first two years of college. CLEP General Examinations are available for English Composition (with or without essay), Humanities, College Mathematics, Natural Sciences, and Social Sciences and History.

2. **CLEP Subject Examinations** include material usually covered in an undergraduate course with a similar title. For a complete list of the subject examinations offered, visit the College Board website.

Who administers the exam?

The CLEP exams are developed by the College Board, administered by Educational Testing Service (ETS), and involve the assistance of educators throughout the United States. The test development process is designed and implemented to ensure that the content and difficulty level of the test are appropriate.

When and where is the exam given?

CLEP exams are administered each month throughout the year at more than 1,300 test centers in the United States and can be arranged for candidates abroad on request. To find the test center nearest you and to register for the exam, you should obtain a copy of the free booklets *CLEP Colleges* and *CLEP Information for Candidates and Registration Form*. They are available at most colleges where CLEP credit is granted, or by contacting:

CLEP Services
P.O. Box 6600
Princeton, NJ 08541-6600
Phone: (800) 257-9558 (8 A.M. to 6 P.M. ET)
Fax: (609) 771-7088
Website: *www.collegeboard.com/clep*

CLEP Options for Military Personnel and Veterans

CLEP exams are available free of charge to eligible military personnel and eligible civilian employees. All the CLEP exams are available at test centers on college campuses and military bases. In addition, the College Board has developed a paper-based version of 14 high-volume/high-pass-rate CLEP tests for DANTES Test Centers. Contact the Educational Services Officer or Navy College Education Specialist for more information. Visit the College Board website for details about CLEP opportunities for military personnel.

Eligible U.S. veterans can claim reimbursement for CLEP exams and administration fees pursuant to provisions of the Veterans Benefits Improvement Act of 2004. For details on eligibility and submitting a claim for reimbursement, visit the U.S. Department of Veterans Affairs website at *www.gibill.va.gov/pamphlets/testing.htm*.

SSD Accommodations for Students with Disabilities

Many students qualify for extra time to take the CLEP Precalculus exam, but you must make these arrangements in advance. For information, contact:

College Board Services for Students with Disabilities
P.O. Box 6226
Princeton, NJ 08541-6226
Phone: (609) 771-7137 (Monday through Friday, 8 A.M. to 6 P.M. ET)
TTY: (609) 882-4118
Fax: (609) 771-7944
E-mail: *ssd@info.collegeboard.org*

HOW TO USE THIS BOOK

What do I study first?

Read over the course review and the suggestions for test-taking, take the first practice test to determine your area(s) of weakness, and then go back and focus your study on those specific problems. Studying the reviews thoroughly will reinforce the basic skills you will need to do well on the exam. Make sure to take the practice tests to become familiar with the format and procedures involved with taking the actual exam.

To best utilize your study time, follow our Independent Study Schedule, which you'll find in the front of this book. The schedule is based on a four-week program, but can be condensed to two weeks if necessary by collapsing each two-week period into one.

When should I start studying?

It is never too early to start studying for the CLEP Precalculus exam. The earlier you begin, the more time you will have to sharpen your skills. Do not procrastinate! Cramming is not an effective way to study, since it does not allow you the time needed to learn the test material. The sooner you learn the format of the exam, the more time you will have to familiarize yourself with it.

FORMAT AND CONTENT OF THE EXAM

The CLEP Precalculus exam covers the material one would find in a college-level introductory calculus class. A majority of the exam tests the test-taker's knowledge and understanding of functions and their properties. Specifically, the following types of functions are tested: linear, quadratic, absolute value, square root, polynomial, rational, exponential, logarithmic, trigonometric, inverse trigonometric, and piecewise-defined.

The exam contains 48 questions in two sections. Section 1 contains 25 questions to be answered in 50 minutes. The use of an online graphing calculator is allowed for this section. Section 2 contains 23 questions to be answered in 40 minutes. Use of a calculator for this section is not allowed.

The approximate breakdown of topics is as follows:

20% Algebraic Expressions, Equations, and Inequalities
15% Functions: Concept, Properties, and Operations
30% Representations of Functions: Symbolic, Graphical, and Tabular
10% Analytic Geometry
15% Trigonometry and its Applications
10% Functions as Models

The online graphing calculator allowed for use in Section 1 of the exam is available to students as a free download for a 30-day trial period. Students are expected to become familiar with the calculator and its functionality prior to the exam. For more information about downloading the practice version of the calculator, visit the Precalculus exam description page on the College Board website at *www.collegeboard.com*.

ABOUT OUR COURSE REVIEW

The review in this book provides you with a complete rundown of all the important mathematical theories and principles relevant to the exam. It will help reinforce the facts you have already learned while better shaping your understanding of the discipline as a whole. By using the review in conjunction with the practice tests, you should be well prepared to take the CLEP Precalculus exam.

SCORING YOUR PRACTICE TESTS

How do I score my practice tests?

The CLEP Precalculus exam is scored on a scale of 20 to 80. To score your practice tests, count up the number of correct answers. This is your total raw score. Convert your raw score to a scaled score using the conversion table on the following page. (*Note:* The conversion table provides only an *estimate* of your scaled score. Scaled scores can and do vary over time, and in no case should a sample test be taken as a precise predictor of test performance. Nonetheless, our scoring table allows you to judge your level of performance within a reasonable scoring range.)

When will I receive my score report?

The test administrator will print out a full Candidate Score Report for you immediately upon your completion of the exam (except for CLEP English Composition with Essay). Your scores are reported only to you, unless you ask to have them sent elsewhere. If you want your scores reported to a college or other institution, you must say so when you take the examination. Since your scores are kept on file for 20 years, you can also request transcripts from Educational Testing Service at a later date.

STUDYING FOR THE CLEP

It is very important for you to choose the time and place for studying that works best for you. Some students may set aside a certain number of hours every morning, while others may choose to study at night before going to sleep. Other students may study during the day, while waiting on a line, or even while eating lunch. Only you can determine when and where your study time will be most effective. But be consistent and use your time wisely. Work out a study routine and stick to it!

PRACTICE TEST RAW SCORE CONVERSION TABLE*

Raw Score	Scaled Score	Course Grade	Raw Score	Scaled Score	Course Grade
48	80	A	23	48	C
47	79	A	22	46	C
46	78	A	21	44	C
45	77	A	20	43	C
44	76	A	19	41	D
43	75	A	18	39	D
42	74	A	17	37	D
41	73	A	16	36	D
40	72	A	15	35	D
39	71	A	14	34	D
38	70	B	13	33	D
37	69	B	12	32	D
36	68	B	11	31	D
35	67	B	10	30	D
34	66	B	9	29	F
33	64	B	8	28	F
32	61	B	7	27	F
31	60	B	6	26	F
30	59	B	5	25	F
29	57	B	4	24	F
28	54	B	3	23	F
27	53	B	2	22	F
26	52	B	1	21	F
25	51	C	0	20	F
24	50	C			

* This table is provided for scoring REA practice tests only. The American Council on Education recommends that colleges use a single across-the-board credit-granting score of 50 for all CLEP computer-based exams. Nonetheless, on account of the different skills being measured and the unique content requirements of each test, the actual number of correct answers needed to reach 50 will vary. A 50 is calibrated to equate with performance that would warrant the grade C in the corresponding introductory college course.

When you take the practice tests, try to make your testing conditions as much like the actual test as possible. Turn your television and radio off, and sit down at a quiet table free from distraction. Make sure to time yourself. Start off by setting a timer for the time that is allotted for each section, and be sure to reset the timer for the appropriate amount of time when you start a new section.

As you complete each practice test, score your test and thoroughly review the explanations to the questions you answered incorrectly; however, do not review too much at one time. Concentrate on one problem area at a time by reviewing the question and explanation, and by studying our review until you are confident that you completely understand the material.

TEST-TAKING TIPS

Although you may not be familiar with computer-based standardized tests such as the CLEP Precalculus exam, there are many ways to acquaint yourself with this type of examination and to help alleviate your test-taking anxieties. Listed below are ways to help you become accustomed to the CLEP, some of which may be applied to other standardized tests as well.

Know the format of the CBT. CLEP CBTs are not adaptive but rather fixed-length tests. In a sense, this makes them kin to the familiar paper-and-pencil exam in that you have the same flexibility to go back and review your work in each section. Moreover, the format isn't a great deal different from the paper-and-pencil CLEP.

Read all of the possible answers. Just because you think you have found the correct response, do not automatically assume that it is the best answer. Read through each choice to be sure that you are not making a mistake by jumping to conclusions.

Use the process of elimination. Go through each answer to a question and eliminate as many of the answer choices as possible. By eliminating just two answer choices, you give yourself a better chance of getting the item correct, since there will be only three choices left from which to make your guess. Remember, your score is based only on the number of questions you answer correctly.

Work quickly and steadily. You will have only 90 minutes to work on 48 questions, so work quickly and steadily to avoid focusing on any one question too long. Taking the practice tests in this book will help you learn to budget your time.

Acquaint yourself with the computer screen. Familiarize yourself with the CLEP computer screen beforehand by logging on to the College Board website. Waiting until test day to see what it looks like in the pretest tutorial risks injecting needless anxiety into your testing experience. Also, familiarizing yourself with the directions and format of the exam will save you valuable time on the day of the actual test.

Be sure that your answer registers before you go to the next item. Look at the screen to see that your mouse-click causes the pointer to darken the proper oval. This takes less effort than darkening an oval on paper, but don't lull yourself into taking less care!

THE DAY OF THE EXAM

On the day of the test, you should wake up early (hopefully after a decent night's rest) and have a good breakfast. Make sure to dress comfortably, so that you are not distracted by being too hot or too cold while taking the test. Also plan to arrive at the test center early. This will allow you to collect your thoughts and relax before the test, and will also spare you the anxiety that comes with being late. As an added incentive to make sure you arrive early, keep in mind that no one will be allowed into the test session after the test has begun.

Before you leave for the test center, make sure that you have your admission form and another form of identification, which must contain a recent photograph, your name, and signature (i.e., driver's license, student identification card, or current alien registration card). You will not be admitted to the test center if you do not have proper identification.

If you would like, you may wear a watch to the test center. However, you may not wear one that makes noise, because it may disturb the other test-takers. No dictionaries, textbooks, notebooks, briefcases, or packages will be permitted and drinking, smoking, and eating are prohibited.

Good luck on the CLEP Precalculus exam!

CHAPTER 2
Algebra Review

Chapter 2
Algebra Review

FACTORING

As you may recall from algebra, factoring is one of the most important math skills you will learn. Factoring can be thought of as the reverse of multiplication. For example, if we multiply $2a(b + 3c)$, we get $2ab + 6ac$; so, if we factor $2ab + 6ac$, we get $2a(b + 3c)$ by factoring out the common factor $2a$. We review some basic types of factoring problems here, but you may want to consult a college algebra text for detailed explanations, especially if you are having difficulty following the examples.

Example

Factor $2a^2 + 4ab - 6b^2$.

Solution

We factor out the common factor, 2:

$2(a^2 + 2ab - 3b^2)$

We can continue by factoring the trinomial into the product of two binomials (the reverse of the FOIL method). Here, the product of the first terms of the binomials gives the first term of the trinomial a^2; the sum of the inner and outer products of the binomials ($+3ab$ and $-ab$) equals the middle term of the trinomial; and the product of the last terms of the binomials equals the last term of the trinomial ($-3b^2$). We then get

$2(a + 3b)(a-b)$

Example

Factor $x^3 - 4x^2 + 2x - 8$ by grouping.

Solution

In this form there are no common factors. But we do notice that there is a common factor (x^2) in the first two terms and a common factor (2) in the last two terms.

$x^2(x - 4) + 2(x - 4)$

We now factor out the common factor $(x - 4)$ in the two remaining terms.

$(x - 4)(x^2 + 2)$

Example

Factor $9x^2 - 16y^2$.

Solution

We recognize this expression as the difference of perfect squares, which can be factored into the product of the sum and difference of the square roots.

$(3x + 4y)(3x - 4y)$

Example

Factor $8a^3 - b^3$.

Solution

This is the difference of perfect cubes. The sum and difference of cubes can be factored into a binomial and a trinomial. We can write $8a^3 - b^3$ as $(2a)^3 - (b)^3$. The binomial is the difference of the cube roots $(2a - b)$ and the trinomial we get from the binomial: square the first term, take the product, square the last term. The sign of the second term in the trinomial is the opposite of the middle sign of the binomial.

$(2a - b)((2a)^2 + (2a)(b) + b^2)$

↑————↑　　　↑

These signs alternate.　　*Last sign is always positive.*

or　$(2a - b)(4a^2 + 2ab + b^2)$

Similarly, $8a^3 + b^3$ would be factored into $(2a + b)(4a^2 - 2ab + b^2)$, i.e., the terms of the binomial $(2a + b)$ are the cube roots of the terms in the given expression, with the same sign. We get the trinomial from the binomial $(2a + b)$ by squaring the first term, taking the product of the two terms (with a sign opposite to the middle sign of the binomial) and then squaring the last term of the binomial.

ALGEBRAIC FRACTIONS

Algebraic fractions are fractions of the form $\frac{p(x)}{q(x)}$, where p and q are polynomials and $q(x) \neq 0$. The rules for addition, subtraction, multiplication, division, and reducing of algebraic fractions follow the same rules as for fractions in arithmetic.

Addition and Subtraction

Following the same rules for fractions of arithmetic, you must have a common denominator in order to add or subtract algebraic fractions.

Example

Add the algebraic fractions $\frac{3a+1}{a-1} + \frac{a+1}{a+5}$.

Solution

We first get a common denominator of $(a - 1)(a + 5)$ and rewrite each of the fractions with the new common denominator, keeping in mind that multiplication by 1 doesn't change each term. In this case, we multiply the first term by $\frac{a+5}{a+5}$ and the second term by $\frac{a-1}{a-1}$ (each equal to 1) to get

$$\frac{(3a+1)(a+5)}{(a-1)(a+5)} + \frac{(a+1)(a-1)}{(a+5)(a-1)}$$

We now multiply out and combine the numerators.

$$\frac{3a^2+16a+5}{(a-1)(a+5)}+\frac{a^2-1}{(a+5)(a-1)}=\frac{4a^2+16a+4}{a^2+4a-5}=\frac{4(a^2+4a+1)}{(a-1)(a+5)}$$

Although we can do some factoring in the numerator and denominator, there will be no common factors that we can cancel, so this is the simplified answer.

Multiplication

Following the rule for multiplying fractions of arithmetic, multiply the numerators and denominators separately. A good general rule is to factor everything so that you can divide out (or cancel) common factors.

Example

Multiply and simplify $\frac{x^2-4}{x^2-1}\times\frac{x^2+2x+1}{x^2-x-2}$.

Solution

We factor the numerators and denominators.

$$\frac{(x-2)(x+2)}{(x-1)(x+1)}\times\frac{(x+1)(x+1)}{(x-2)(x+1)}$$

We now cancel common factors in the numerator and denominator to get $\frac{x+2}{x-1}$.

Division

Use the general rule from fractions of arithmetic: "invert and multiply." Inverting means using the reciprocal of the divisor.

Example

Divide and simplify $\frac{x^2-9}{x^2-9x+18}\div\frac{x^2+5x+6}{x^2-x-30}$.

Solution

We begin by factoring all numerators and denominators.

$$\frac{(x-3)(x+3)}{(x-3)(x-6)} \div \frac{(x+2)(x+3)}{(x-6)(x+5)}$$

We now invert the divisor fraction and multiply.

$$\frac{(x-3)(x+3)}{(x-3)(x-6)} \times \frac{(x-6)(x+5)}{(x+2)(x+3)}$$

We now cancel common factors to simplify: $\frac{x+5}{x+2}$.

Complex Fractions

Complex fractions have fractions within fractions.

Example

Simplify the complex fraction $\frac{1}{\frac{1}{a+b}+\frac{1}{a-b}}$.

Solution

We multiply both the numerator and denominator of the complex fraction by common factors that will clear out the fractions within the fraction. In this problem, we multiply both the numerator and denominator by $(a + b)(a - b)$, remembering to multiply each term of the denominator.

$$\left(\frac{(a+b)(a-b)}{(a+b)(a-b)}\right) \times \left(\frac{1}{\frac{1}{a+b}+\frac{1}{a-b}}\right)$$

Notice that the denominator simplifies as you multiply. This gives us

$$\frac{(a+b)(a-b)}{(a-b)+(a+b)}$$

which simplifies to

$$\frac{a^2 - b^2}{2a}$$

Therefore,

$$\frac{1}{\dfrac{1}{a+b} + \dfrac{1}{a-b}} = \frac{a^2 - b^2}{2a}$$

We can factor the numerator if we wish, but there are no common factors that will cancel, so this is the simplified answer.

Example

Simplify the complex fraction $\dfrac{1 + \dfrac{1}{a}}{1 - \dfrac{a}{3}}$.

Solution

We begin by multiplying the numerator and denominator by the common denominator $3a$.

$$\left(\frac{3a}{3a}\right) \times \left(\frac{1 + \dfrac{1}{a}}{1 - \dfrac{a}{3}}\right)$$

We now have

$$\frac{3a + 3}{3a - a^2}$$

Hence,

$$\frac{1 + \dfrac{1}{a}}{1 - \dfrac{a}{3}}$$

which simplifies to

$$\frac{3a + 3}{3a - a^2}$$

Although we can factor the numerator and denominator, again there are no common factors we can cancel.

SOLVING LINEAR AND QUADRATIC EQUATIONS

Solving Linear Equations

To solve $ax + b = c$ for x, we "isolate" the x on one side of the equals sign by subtracting b from both sides of the equation and then dividing by the coefficient of x.

$$ax + b = c$$

$$ax = c - b$$

$$x = \frac{c - b}{a}$$

By using this process, along with our operations of algebra, we can solve any type of linear equation.

Example

Solve for P in the formula $A = P + Prt$.

Solution

We begin by factoring P out of both terms on the right side of the equation.

$$A = P(1 + rt)$$

Dividing both sides by $(1 + rt)$ to isolate P, we have

$$\frac{A}{1 + rt} = P$$

Example

Solve for c_1 in the formula $\frac{1}{c} = \frac{1}{c_1} + \frac{1}{c_2}$.

Solution

We clear out fractions by multiplying by the factors $c \times c_1 \times c_2$.

$$(c \times c_1 \times c_2)\left(\frac{1}{c} = \frac{1}{c_1} + \frac{1}{c_2}\right)$$

$c_1c_2 = cc_2 + cc_1$

Putting the terms with c_1 on one side of the equation, we get

$c_1c_2 - cc_1 = cc_2$

Then, by factoring and dividing, we get

$c_1(c_2 - c) = cc_2$

$$c_1 = \frac{cc_2}{c_2 - c}$$

Example

Solve for R in the formula $I = \dfrac{E}{R+r}$.

Solution

First, we clear out the fraction by multiplying both sides of the equation by $(R + r)$ and then solve for R.

$I(R + r) = E$

$IR + Ir = E$

$IR = E - Ir$

$$R = \frac{E - Ir}{I}$$

Solving Quadratic Equations

The general quadratic equation is of the form $ax^2 + bx + c = 0$. There are several methods we can use to solve quadratic equations, or equations we can put into quadratic form.

Factoring

If a quadratic equation can be factored, either factor must equal zero.

Example

Solve $6x^2 - 17x + 5 = 0$ for x by factoring.

Solution

$(3x - 1)(2x - 5) = 0$

We set each factor equal to 0 and solve for x.

$3x - 1 = 0 \qquad 2x - 5 = 0$

$x = \frac{1}{3} \qquad x = \frac{5}{2}$

Completing the Square

Another process that we can use to solve quadratic equations is completing the square. In this process, we use the fact that if you add the same quantity to both sides of an equation, the two sides are still equal. We want to create a perfect trinomial, which is the square of a binomial on one side of the equation, and then take the square root of each side to solve for the unknown.

Note: In general, the steps in the process of completing the square for all quadratics are:

1. If the coefficient of x^2 is not $+1$, then divide everything by the coefficient of x^2.
2. Place all the terms involving x on one side of the equal sign and the constant on the other side.
3. Take one-half of the coefficient of x.
4. Square this number and add it to both sides of the equation.
5. You have created a perfect trinomial in x. Write the perfect trinomial as a binomial squared.
6. If solving for x, take the square root of both sides and solve, simplifying if necessary.

Example

Solve $x^2 - 6x + 1 = 0$ for x by completing the square.

Solution

We complete the square on the variable x to create the perfect trinomial by taking one-half the coefficient of x (which is half of (-6) or (-3)), and then squaring it and adding it to both sides. To use this process of completing the square in solving this equation, we have

$$x^2 - 6x = -1$$

$$x^2 - 6x + \underline{\quad} = -1 + \underline{\quad}$$

$$x^2 - 6x + \underline{9} = -1 + \underline{9}$$

$$(x - 3)^2 = 8$$

$$\sqrt{(x - 3)^2} = \pm\sqrt{8}$$

$$x - 3 = \pm\sqrt{8}$$

$$x = 3 \pm \sqrt{8}$$

We can simplify $\sqrt{8}$ so that our answer becomes

$$x = 3 \pm 2\sqrt{2}$$

Example

Complete the square on the trinomial $x^2 - 10x - 4 = 0$ to solve for x.

Solution

We follow the steps listed above.

1. The coefficient of x^2 is equal to $+1$, so we can skip step 1.
2. $x^2 - 10x = 4$
3. The coefficient of x is -10, so
   $\frac{1}{2}$ of (-10) is (-5)
4. We square -5 and add it to both sides of the equation:
   $x^2 - 10x + (-5)^2 = 4 + (-5)^2$
   or, by simplifying:
   $x^2 - 10x + 25 = 4 + 25$
5. The left side is a perfect trinomial in x. So we have
   $(x - 5)^2 = 29$

6. Taking the square root of both sides of the resulting equation, we get

$$\sqrt{(x-5)^2} = \pm\sqrt{29}$$
$$(x-5) = \pm\sqrt{29}$$
$$x = 5 \pm \sqrt{29}$$

Example

Solve $2x^2 + 3x - 4 = 0$ for x by completing the square.

Solution

The process of completing the square requires that the coefficient of x^2 is equal to 1. Therefore, we must first divide everything by 2 and then continue with the process.

$$x^2 + \frac{3}{2}x - 2 = 0$$

We put the constant on the other side of the equal sign.

$$x^2 + \frac{3}{2}x + \_\_\_ = 2 + \_\_\_$$

Now we figure what to add to both sides. We take $\frac{1}{2}$ of the coefficient of x, or $\left(\frac{1}{2}\right) \times \left(\frac{3}{2}\right)$, which is $\frac{3}{4}$. We now square this number and add it to both sides of the equation.

$$x^2 + \frac{3}{2}x + \frac{9}{16} = 2 + \frac{9}{16}$$

We write the perfect trinomial on the left side of the equation as a binomial squared.

$$\left(x + \frac{3}{4}\right)^2 = \frac{41}{16}$$

We take the square root of both sides and solve for x.

$$\sqrt{\left(x+\frac{3}{4}\right)^2} = \pm\sqrt{\frac{41}{16}}$$

$$\left(x+\frac{3}{4}\right) = \pm\frac{\sqrt{41}}{4}$$

$$x = -\frac{3}{4} \pm \frac{\sqrt{41}}{4}, \text{ or } \frac{-3 \pm \sqrt{41}}{4}$$

Quadratic Formula

The process of completing the square on the general quadratic equation $ax^2 + bx + c = 0$ can be used to solve for x and gives us the quadratic formula:

$$x = \frac{-b \pm \sqrt{b^2 - 4ac}}{2a}$$

The expression $b^2 - 4ac$ is called the discriminant and tells us something about the solutions to the quadratic equation $ax^2 + bx + c = 0$. We will explore the situation when the discriminant is negative when we discuss complex numbers in the next section.

Let's redo the equation in the last example to see how the quadratic formula works.

$2x^2 + 3x - 4 = 0$

$a = 2$, $b = +3$, and $c = -4$. Then

$$x = \frac{-3 \pm \sqrt{(3)^2 - 4(2)(-4)}}{2(2)} = \frac{-3 \pm \sqrt{41}}{4}$$

Example

Solve $3x^2 - 2x - 4 = 0$ for x by using the quadratic formula.

Solution

Here $a = 3$, $b = -2$, and $c = -4$; we substitute these values in the formula to obtain

$$x = \frac{2 \pm \sqrt{(-2)^2 - 4(3)(-4)}}{2(3)} = \frac{2 \pm \sqrt{52}}{6} = \frac{1 \pm \sqrt{13}}{3}$$

Putting Equations into Quadratic Form

Sometimes we can put equations that are not quadratic into a quadratic form and then use the techniques of solving quadratic equations.

Example

Solve $x^4 - 5x^2 + 4 = 0$ for x.

Solution

This is a quadratic equation in x^2 so we can use one of our procedures to solve for x^2 and then for x. By factoring (here x^2 is treated the same as the "x" term in our standard form), we have

$$(x^2 - 4)(x^2 - 1) = 0$$

or solving each for x gives us $x = \pm 2$ or $x = \pm 1$.

Example

Solve $y^6 - 3y^3 - 2 = 0$ for y.

Solution

This is a quadratic in y^3. Using the quadratic formula with $a = 1$, $b = -3$, $c = -2$ (our usual x is y^3 here), we get

$$y^3 = \frac{3 \pm \sqrt{(-3)^2 - 4(1)(-2)}}{2(1)}$$

$$y^3 = \frac{3 \pm \sqrt{17}}{2}$$

$$y = \sqrt[3]{\frac{3 \pm \sqrt{17}}{2}}$$

Example

Solve $5x^{\frac{2}{3}} - 7x^{\frac{1}{3}} = 6$ for x.

Solution

This is a quadratic in $x^{\frac{1}{3}}$ that we must set equal to 0, or $5x^{\frac{2}{3}} - 7x^{\frac{1}{3}} - 6 = 0$. Using the quadratic formula and $a = 5$, $b = -7$, and $c = -6$, we get

$$x^{\frac{1}{3}} = \frac{7 \pm \sqrt{(-7)^2 - 4(5)(-6)}}{2(5)}$$

$$x^{\frac{1}{3}} = \frac{7 \pm \sqrt{169}}{10} = \frac{7 \pm 13}{10}$$

$$x^{\frac{1}{3}} = \frac{7 + 13}{10} = \frac{20}{10} \quad \text{or} \quad x^{\frac{1}{3}} = \frac{7 - 13}{10} = -\frac{6}{10}$$

$$x^{\frac{1}{3}} = 2 \quad \text{or} \quad x^{\frac{1}{3}} = -\frac{3}{5}$$

$$\left(x^{\frac{1}{3}}\right)^3 = (2)^3 \quad \text{or} \quad \left(x^{\frac{1}{3}}\right)^3 = \left(-\frac{3}{5}\right)^3$$

$$x = 8 \quad \text{or} \quad x = -\frac{27}{125}$$

COMPLEX NUMBERS

Numbers of the form $a + bi$, where $i = \sqrt{-1}$ and a, b are real numbers, are called complex numbers. The term containing i is the imaginary term. So far, our examples have all had discriminants of $b^2 - 4ac \geq 0$ so that our answers have all been real numbers. If $b^2 - 4ac < 0$, then the solutions to the quadratic equations will be complex numbers. We now review the basics of complex numbers, which can be simplified by using the following facts.

$$\begin{array}{lll}
i=\sqrt{-1} & i^5 = i^4 \times i = i & i^9 = (i^4)^2 \times i = i \\
i^2 = -1 & i^6 = i^4 \times i^2 = -1 & i^{10} = (i^4)^2 \times i^2 = -1 \\
i^3 = i^2 \times i = -i & i^7 = i^4 \times i^3 = -i & i^{11} = (i^4)^2 \times i^3 = -i \\
i^4 = i^2 \times i^2 = 1 & i^8 = i^4 \times i^4 = 1 & i^{12} = (i^4)^3 = 1
\end{array}$$

This pattern repeats to allow us to simplify any power of i to one of the following: i, -1, $-i$, 1.

Example

Simplify (a) i^{121} and (b) i^{63}.

Solution

Since the pattern repeats every four powers of i, we divide the exponent of i by 4, and the remainder indicates the value: $i^1 = 1$, $i^2 = -1$, $i^3 = -1$, $i^4 = 1$.

(a) $i^{121} = (i^4)^{30} \times i = (1)^{30} \times i = 1 \times i = i$

(b) $i^{63} = (i^4)^{15} \times i^3 = (1)^{15} \times i^3 = 1 \times (-i) = -i$

Operations of Complex Numbers

The operations of complex numbers are similar to those of real numbers. To add or subtract, we combine the real parts and then the imaginary parts of the complex number.

Example

Combine: (a) $(1 + 2i) + (5 - i)$ and (b) $(3 - i) - (2 - 7i)$

Solution

(a) $(1 + 2i) + (5 - i) = 6 + i$

(b) $(3 - i) - (2 - 7i) = (3 - i) + (-2 + 7i) = 1 + 6i$

The rules for multiplying and dividing complex numbers are similar to those of real number arithmetic, except we simplify any terms with $i^2 = -1$, and we put the answer into complex form, $a + bi$. Multiplying, we can use FOIL and then simplify.

Example

Multiply: (a) $(2 - i)(3 + 5i)$ and (b) $(4 + 5i)(4 - 5i)$

Solution

(a) $(2 - i)(3 + 5i) = 6 + 10i - 3i - 5i^2 = 6 + 7i + 5 = 11 + 7i$

(b) $(4 + 5i)(4 - 5i) = 16 - 20i + 20i - 25i^2 = 16 + 25 = 41$

To discuss division of complex numbers, we first must define the *conjugate* of a complex number. In Example (b) above, we notice that when we multiply $(4 + 5i)$ and $(4 - 5i)$, we obtain a real number for their product since the imaginary terms combine to equal 0. These factors are called conjugates. Notice that conjugates have the same first terms and opposite second terms. For example, the conjugate of $(-3 + 7i)$ is $(-3 - 7i)$—only the second term sign is changed. To divide complex numbers, we multiply both the numerator and the denominator by the conjugate of the denominator, thus creating a real number in the denominator. We write the final answer in standard form $a + bi$.

Example

Divide $\frac{2+3i}{6-i}$.

Solution

To divide two complex numbers, we multiply both the numerator and the denominator by the conjugate of the denominator.

$$\frac{2+3i}{6-i} = \frac{(2+3i)(6+i)}{(6-i)(6+i)} = \frac{12+20i+3i^2}{36-i^2} = \frac{9+20i}{37} = \frac{9}{37} + \frac{20}{37}i$$

Complex Roots of Quadratic Equations

We now continue with a quadratic equation that has complex roots.

Example

Solve $2x^2 + x + 4 = 0$.

Solution

We use $a = 2$, $b = 1$, and $c = 4$ in the quadratic formula to obtain

$$x = \frac{-1 \pm \sqrt{(1)^2 - 4(2)(4)}}{2(2)} = \frac{-1 \pm \sqrt{-31}}{4} = \frac{-1 \pm i\sqrt{31}}{4} = -\frac{1}{4} \pm \frac{\sqrt{31}}{4}i$$

Note: This quadratic equation has only complex solutions, which means that the graph of the quadratic function $f(x) = 2x^2 + x + 4$ does not cross the x-axis. There are no real values of x for which $2x^2 + x + 4 = 0$.

SOLVING RADICAL EQUATIONS

Radical equations are equations in which the variable appears under a radical. These equations present the possibilities of obtaining "extraneous" roots, or solutions that do not check in the equation. To solve radical equations, we must eliminate the radicals by raising to a power. We first isolate the radical expression on one side of the equation.

Example

Solve $\sqrt{2-x} - x = 0$.

Solution

$\sqrt{2-x} = x$

We square both sides to clear out the radical.

$(\sqrt{2-x})^2 = x^2$

$2 - x = x^2$

We set the equation equal to 0 and solve.

$x^2 + x - 2 = 0$

$(x + 2)(x - 1) = 0$

$x = -2 \quad \text{or} \quad x = 1$

The only correct solution is 1; however, since -2 does not check in the equation—it is an extraneous solution.

Anytime we square (or raise to any even power), we introduce the possibility of extraneous answers. We must always check answers back in the original equation. It is a good idea to do this for any equation, but it is necessary for radical equations.

Example

Solve $\sqrt{x+10} + 2 = x$.

Solution

We begin by isolating the radical expression and then squaring both sides (even though we introduce the possibility of extraneous solutions).

$$(\sqrt{x+10})^2 = (x-2)^2$$

$$x + 10 = x^2 - 4x + 4$$

$$x^2 - 5x - 6 = 0$$

$$(x - 6)(x + 1) = 0$$

$$x = 6 \text{ or } x = -1$$

The only correct solution is 6 since -1 is an extraneous solution because it does not check in the equation, $\sqrt{x+10} + 2 = x$.

Example

Solve for $\sqrt{2x+3} - \sqrt{x+1} = 1$.

Solution

We first separate the radicals by placing one on each side of the equal sign.

$$\sqrt{2x+3} = 1 + \sqrt{x+1}$$

We now square both sides to get

$$(\sqrt{2x+3})^2 = (1 + \sqrt{x+1})^2$$

$$2x + 3 = 1 + 2\sqrt{x+1} + (x + 1)$$

When we combine terms, we still have a radical term, so we isolate the new radical expression, square both sides, and solve for x.

$$x + 1 = 2\sqrt{x+1}$$

$$(x+1)^2 = (2\sqrt{x+1})^2$$

$$x^2 + 2x + 1 = 4(x + 1)$$

$$x^2 - 2x - 3 = 0$$

$$(x - 3)(x + 1) = 0$$

$$x = 3 \text{ or } x = -1$$

We now must check the answers into the *original* equation. Both check to be solutions here.

SOLVING EQUATIONS WITH RATIONAL EXPONENTS

To solve equations with rational exponents, we must raise both sides of the equation to a power that is the reciprocal of the rational exponents to make the rational exponent a 1. This will allow us to solve for x.

Example

Solve $(x-2)^{\frac{2}{3}} = 9$.

Solution

The reciprocal of the rational exponent $\frac{2}{3}$ is $\frac{3}{2}$. Therefore, we raise both sides of the equation to the power $\frac{3}{2}$.

$$\left((x-2)^{\frac{2}{3}}\right)^{\frac{3}{2}} = 9^{\frac{3}{2}} = (\sqrt{9})^3 = 3^3$$
$$(x-2)^1 = 27$$
$$x - 2 = 27$$
$$x = 29$$

This number checks in the original equation.

Example

Solve $(3x+1)^{\frac{1}{3}} = (4x+9)^{\frac{1}{3}}$.

Solution

To solve, we cube each side of the equation, since the reciprocal of $\frac{1}{3}$ is 3.

$$\left((3x+1)^{\frac{1}{3}}\right)^3 = \left((4x+9)^{\frac{1}{3}}\right)^3$$
$$(3x+1)^1 = (4x+9)^1$$
$$3x + 1 = 4x + 9$$
$$x = -8$$

This answer checks in the original equation.

SOLVING RATIONAL EQUATIONS

A rational equation is an equation with fractions. The best rule to use in solving rational equations is to clear out all the fractions. We do so by multiplying both sides of the equation by an expression all the factors in the denominators will divide into evenly. This is called the lowest common multiple (LCM). For example, to solve

$$\frac{3}{x} + \frac{2x-5}{x} = 6 \qquad x \neq 0$$

we multiply everything by x to clear the fractions.

$$x \times \left(\frac{3}{x} + \frac{2x-5}{x} = 6\right)$$

$$3 + (2x - 5) = 6x$$

Note: We multiplied every term by x, even the 6.

$$2x - 2 = 6x$$

$$-4x = 2$$

$$x = -\frac{1}{2}$$

We must be sure to check our answer!

Example

Solve $\dfrac{1}{x+2} + \dfrac{2}{x-3} = \dfrac{8}{x^2 - x - 6}$ for x $\quad (x \neq -2, x \neq 3)$.

Solution

We first notice that $x^2 - x - 6$ can be factored into the product $(x + 2)(x - 3)$.

$$\frac{1}{x+2} + \frac{2}{x-3} = \frac{8}{(x+2)(x-3)}$$

We now multiply everything by $(x + 2)(x - 3)$, since each denominator is a factor of this expression, to obtain

$$(x+2)(x-3)\left(\frac{1}{x+2} + \frac{2}{x-3} = \frac{8}{(x+2)(x-3)}\right)$$

$$1(x - 3) + 2(x + 2) = 8$$

Solving, we get $x = \frac{7}{3}$, which checks in the equation.

SOLVING ABSOLUTE VALUE EQUATIONS

We recall the defintion of absolute value:

$|x| = x$, if $x \geq 0$

$|x| = -x$, if $x < 0$

We use this definition to solve absolute value equations. Note that the solution will generally have two choices.

Example

Solve $|x + 1| = 9$.

Solution

$x + 1 = 9$ or $x + 1 = -9$

$x = 8$ or $x = -10$

Example

Solve $|2 - y| = 3$.

Solution

$2 - y = 3$ or $2 - y = -3$

$y = -1$ or $y = 5$

We must be sure and check our answers!

Example

Solve $|x - 7| = |3x + 9|$ for x.

Solution

$x - 7 = 3x + 9$ or $x - 7 = -(3x + 9)$

$-2x = 16$ or $4x = -2$

$x = -8$ or $x = -\frac{1}{2}$

These answers check in the original equation.

Example

Solve $\left|\frac{3x-7}{2x+1}\right| = 2$ for x.

Solution

$\frac{3x-7}{2x+1} = 2$ or $\frac{3x-7}{2x+1} = -2$

$3x - 7 = 2(2x + 1)$ or $3x - 7 = -2(2x + 1)$

$3x - 7 = 4x + 2$ or $3x - 7 = -4x - 2$

$-x = 9$ or $7x = 5$

$x = -9$ or $x = \frac{5}{7}$

Both of these answers check in the original equation.

SOLVING INEQUALITIES

Solving inequalities is similar to solving equations, except that the solution is often represented by intervals. Intervals show a beginning value and an ending value, and the inequality symbols tell us whether that value is part of the interval. We use a square bracket to include the endpoint and a parenthesis to exclude it.

The process of solving inequalities is affected by the following rule of inequalities: To multiply or divide an expression by a negative factor reverses the inequality symbol. For example: If $a < b$, then $6a < 6b$, but $-6a > -6b$ since we multiplied both sides by a negative factor, -1, which reversed the inequality sign.

Example

Solve $x + 1 > 4x - 7$ for x.

$8 > 3x$

$x < 8/3$ $(-\infty, 8/3)$

Solution

Just as we do for equations, we first gather like terms and then solve.

$$x + 1 > 4x - 7$$

$$-3x > -8$$

$$x < \frac{8}{3}$$

The inequality symbol is reversed since we divided by (-3). The answer in interval notation is $\left(-\infty, \frac{8}{3}\right)$.

Example

Solve $\frac{5(x-1)}{2} - \frac{3x}{5} \leq \frac{7x-2}{10}$.

25(x-1) - 2(3x) ≤ 7x-2
25x -25 -6x -7x+2 ≤ 0
12x ≤ 23   x ≤ 23/12

Solution

We first clear out the fractions by multiplying by the LCM, which is 10.

$$10\left(\frac{5(x-1)}{2} - \frac{3x}{5}\right) \leq 10\left(\frac{7x-2}{10}\right)$$

$$25(x-1) - 6x \leq 7x - 2$$

Multiplying and combining like terms gives us

$$19x - 25 \leq 7x - 2$$

$$12x \leq 23$$

$$x \leq \frac{23}{12}$$

Interval notation for this answer is $\left(-\infty, \frac{23}{12}\right]$. Note that we used a square bracket to indicate the inclusion of $\frac{23}{12}$ in the answer.

Solving Compound Inequalities

Compound inequalities have either the union (OR) of solutions or the intersection (AND) of solutions. For example, $3 \leq x + 5 \leq 8$ is an example of an intersection. We need values that satisfy both $3 \leq x + 5$ (which we can also write as $x + 5 \geq 3$) *and* $x + 5 \leq 8$.

We can solve each inequality separately, or we can solve as follows:

$$\begin{array}{rcccl} 3 & \leq & x+5 & \leq & 8 \\ -5 & & -5 & & -5 \\ \hline -2 & \leq & x & \leq & 3 \end{array}$$

This gives us $x \geq -2$ and $x \leq 3$, and we can also write the solution as $[-2, 3]$.

Example

Solve $\frac{5x+2}{3} \leq 4$ or $\frac{1-2x}{5} \geq 1$ for x.

Solution

We solve each part of this compound inequality separately and then combine the solutions. The solution set is the *union* of solutions, indicated by the word "or."

$$\begin{array}{lcl} \frac{5x+2}{3} \leq 4 & & \frac{1-2x}{5} \geq 1 \\ 5x + 2 \leq 12 & & 1 - 2x \geq 5 \\ x \leq 2 & & x \leq -2 \\ (-\infty, 2] & \cup & (-\infty, -2] \end{array}$$

Hence, the solution (union) is $(-\infty, 2]$ because this interval includes both answers, and we are looking for one OR the other.

Example

Solve $-2 < \dfrac{3x+1}{7} < 10$ for x.

Solution

We can solve this compound inequality by rewriting it without the fractions and then solving. We first multiply by 7 to clear out fractions and then subtract 1.

$-14 < 3x + 1 < 70$

$\underline{-1 \qquad\quad -1 \quad -1}$

$-15 < 3x < 69$

Dividing by 3, we get

$-5 < x < 23$

Interval notation of this solution is $(-5, 23)$.

Solving Quadratic Inequalities

We must remember two rules when working with quadratic inequalities with a positive coefficient for x^2:

1. less than ($<$) always means *intersection*
2. greater than ($>$) always means *union*

Example

Solve $2x^2 < 15 + x$ for x.

Solution

We solve this inequality similar to how we solve a quadratic equation. Our first step in the solution is to place all terms on one side of the inequality symbol and then factor.

$2x^2 - x - 15 < 0$

$(2x + 5)(x - 3) < 0$

One way to solve this inequality is to find the values of x that make the quadratic equal to 0, or $-\frac{5}{2}$ and 3. Therefore, we have $-\frac{5}{2} < x < 3$ or, using interval notation, $x = \left(-\frac{5}{2}, 3\right)$. The endpoints are excluded. Also, the interval includes only points that satisfy both $-\frac{5}{2} < x$ AND $x < 3$ because it is the intersection. Graphically, we have the configuration in Figure 2.1.

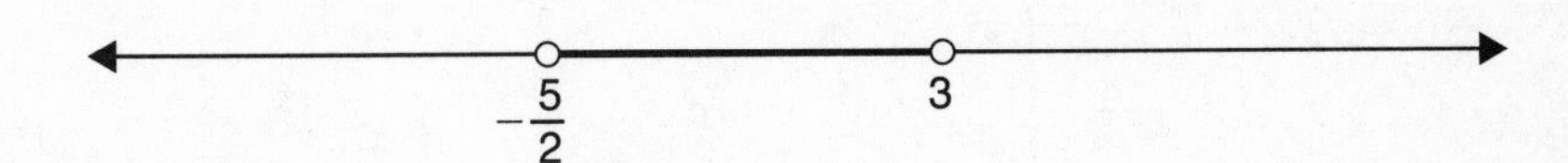

Figure 2.1

Example

Solve $3x^2 - x - 4 \geq 0$ for x.

Solution

We factor (or use the quadratic formula) to solve for the values of x that will make the quadratic equal to 0.

$(3x - 4)(x + 1) = 0$

$3x - 4 = 0 \qquad x + 1 = 0$

$x = \frac{4}{3} \qquad x = -1$

We start at these values that make the quadratic equal to 0 and "go out and back" along the number line, since this is a union (OR). In this case, the endpoints are included.

$x \geq \frac{4}{3} \cup x \leq -1$ or, using interval notation, $x = (-\infty, -1] \cup \left[\frac{4}{3}, \infty\right)$.

The graph is shown in Figure 2.2.

Figure 2.2

Example

Solve $2x - 1 \leq x^2$ for x.

Solution

We first rewrite the inequality in quadratic form and then factor.

$$x^2 - 2x + 1 \geq 0$$

$$(x - 1)(x - 1) \geq 0$$

$$(x - 1)^2 \geq 0$$

This is true for all x, so the solution is $(-\infty, \infty)$.

Solving Rational Inequalities

If the unknown is in the denominator of a rational inequality, we cannot clear the inequality of fractions by just multiplying through by the denominator because that may introduce errors. We must simplify the inequality so that one side is 0, and then consider how the numerator and denominator of the resulting fraction on the other side will make the statement true.

Example

Solve $\dfrac{3x+4}{x+1} \leq 2$ for x.

Solution

We need to combine the terms by first bringing the 2 to the left side of the inequality.

$$\frac{3x+4}{x+1}-2\leq 0$$

Finding a common denominator and simplifying, we have

$$\frac{3x+4}{x+1}-\frac{2(x+1)}{(x+1)}\leq 0$$

$$\frac{3x+4-2(x+1)}{x+1}\leq 0$$

$$\frac{x+2}{x+1}\leq 0$$

For this statement to be true, the fraction must be negative, so either the numerator or the denominator (but not both) must be negative. The only way for the fraction to equal 0 is for the numerator to equal 0.

Case 1		Case 2
$x + 2 \leq 0$ and $x + 1 > 0$	or	$x + 2 \geq 0$ and $x + 1 < 0$
$x \leq -2$ and $x > -1$		$x \geq -2$ and $x < -1$
ϕ		$[-2, -1)$

Therefore, the solution to the problem is $[-2, -1)$.

Example

Solve $\dfrac{x^2+x-6}{x-4} \geq 0$ for x.

Solution

We first factor the numerator.

$$\frac{(x+3)(x-2)}{x-4}\geq 0$$

Since the fraction is nonnegative, the numerator and denominator must have the same sign; for the fraction to equal 0, the numerator must equal 0.

Case 1		Case 2
$(x + 3)(x - 2) \geq 0$ and $x - 4 > 0$	or	$(x + 3)(x - 2) \leq 0$ and $x - 4 < 0$
		$[-3, \infty) \cap [-\infty, 2] \cap x < 4$
$[-3, \infty) \cap [2, \infty) \cap x > 4$		$[-3, 2] \cap x < 4$
$[2, \infty) \cap x > 4$		
$x > 4$		$[-3, 2]$

The final solution is the union of these solutions: $[-3, 2] \cup (4, \infty)$.

Solving Absolute Value Inequalities

The solution of absolute value inequalities is based on the rules of the absolute value. Absolute value inequalities can be separated into two statements, and the solution is either the intersection (<) or union (>) of the solutions of these statements.

Example

Solve $|x - 1| < 3$ for x.

Solution

One of two conditions could be true:

$x - 1 < 3$ or $x - 1 > -3$

We solve each inequality separately, which gives us

$x < 4$ or $x > -2$

The solution is the intersection of the two solutions, $-2 < x < 4$ or, using interval notation, $(-2, 4)$.

Example

Solve $|x - 6| \geq 1$ for x.

Solution

Our two possible solution sets are represented here by

$x - 6 \geq 1$ or $x - 6 \leq -1$

The solution sets are $x \geq 7$ or $x \leq 5$. The solution set is the union of these sets or, in interval notation, $(-\infty, 5] \cup [7, \infty)$.

Example

Solve $|5 - 10x| > -1$ for x.

Solution

By inspection, we can see that all real numbers are solutions to this inequality, since absolute value is always greater than a negative value.

Example

Solve $\frac{1}{4}|3x - 1| - 2 \leq 4$ for x.

Solution

After we add 2 to both sides, we then multiply by 4 to isolate the absolute value expression.

$|3x - 1| \leq 24$

This is a compound inequality that we separate into two statements.

$3x - 1 \leq 24$ or $3x - 1 \geq -24$

Solving each inequality, we get

$$x \leq \frac{25}{3} \text{ or } x \geq -\frac{23}{3}$$

The solution can be represented as $-\frac{23}{3} \leq x \leq \frac{25}{3}$ or, in interval notation, $\left[-\frac{23}{3}, \frac{25}{3}\right]$.

SOLVING SYSTEMS OF EQUATIONS

The solution of a system of equations is a solution to every equation in the system. Since equals added (or subtracted) to equals are equal, the solution can be found by adding (or subtracting) the equations in the system to eliminate one variable, or by substituting from one equation into another, or graphically by finding the intersection of the graphs of the equations, or by a combination of any of these methods.

Example

Solve the following system of linear equations.

$$\frac{x}{3} - y = 2$$

$$x + y = 4(1 - y)$$

Solution

We first multiply the top equation by 3 and simplify the second equation.

$$3\left(\frac{x}{3} - y = 2\right) \Rightarrow \quad x - 3y = 6$$

$$x + y = 4 - 4y \Rightarrow \quad x + 5y = 4$$

We now subtract the second equation from the first.

$$x - 3y = 6$$

$$-x - 5y = -4$$

which yields $-8y = 2$ or $y = -\frac{1}{4}$. Substituting this value into either of the original equations gives us $x = \frac{21}{4}$. Hence, the solution is $(x, y) = \left(\frac{21}{4}, -\frac{1}{4}\right)$. These values of x and y check in both of the original equations.

Example

Solve the following system of equations.

$$x^2 + y^2 = 4$$

$$x + y = 1$$

Solution

Graphically, we are looking for the intersection of a line and a circle. Lines and circles do not have to intersect, so we first graph to see how many points of intersection: 0, 1, or 2. We see from Figure 2.3 that there are two points of intersection.

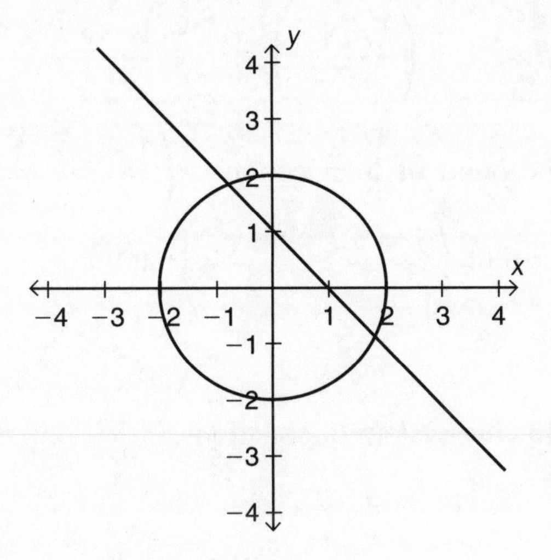

Figure 2.3

We can solve by substitution to get the points of intersection. To begin, we solve the line for y: $y = 1 - x$. We now substitute this value for y into the equation of the circle.

$$x^2 + (1 - x)^2 = 4$$

$$x^2 + (1 - 2x + x^2) = 4$$

$$2x^2 - 2x - 3 = 0$$

We use the quadratic formula to solve.

$$x = \frac{2 \pm \sqrt{(-2)^2 - 4(2)(-3)}}{2(2)} = \frac{2 \pm \sqrt{28}}{4} = \frac{1 \pm \sqrt{7}}{2}$$

This gives us the two x values of the coordinates of the points of intersection. We now substitute each of these values into one of the original equations.

$$x + y = 1$$

$$\left(\frac{1 + \sqrt{7}}{2}\right) + y = 1$$

$$y = 1 - \left(\frac{1 + \sqrt{7}}{2}\right) = \frac{2 - (1 + \sqrt{7})}{2} = \frac{1 - \sqrt{7}}{2}$$

Therefore, one point of intersection is $\left(\frac{1 + \sqrt{7}}{2}, \frac{1 - \sqrt{7}}{2}\right)$. We can find the other point, $\left(\frac{1 - \sqrt{7}}{2}, \frac{1 + \sqrt{7}}{2}\right)$, similarly.

Example

Solve the following system of equations.

$$\frac{2}{x} + \frac{1}{y} = \frac{13}{15}$$

$$\frac{1}{x} - \frac{2}{y} = \frac{4}{15}$$

Solution

We first multiply the top equation by 2 so we can combine it with the second equation.

$$2\left(\frac{2}{x} + \frac{1}{y} = \frac{13}{15}\right) \Rightarrow \frac{4}{x} + \frac{2}{y} = \frac{26}{15}$$

$$\frac{1}{x} - \frac{2}{y} = \frac{4}{15}$$

We now add the two equations together, and the y terms cancel out. We get

$$\frac{5}{x} = \frac{30}{15} = 2$$

Solving for x, we get

$$x = \frac{5}{2}$$

Substituting this value into one of the original equations, we get

$$\frac{1}{\left(\frac{5}{2}\right)} - \frac{2}{y} = \frac{4}{15}$$

$$\frac{2}{5} - \frac{2}{y} = \frac{4}{15}$$

$$-\frac{2}{y} = -\frac{2}{15}$$

$$y = 15$$

The solution is $(x, y) = \left(\frac{5}{2}, 15\right)$. We must be sure to check the values of x and y in both of the original equations.

CHAPTER 3
Functions

Chapter 3
Functions

A *function* $y = f(x)$ is a rule that assigns to each value of x a unique value for y. Graphically, this means that a vertical line should intersect the graph of the function in only one point. The *domain* of the function is the set of all values for x and the *range* is the set of all values for y or $f(x)$.

DOMAIN AND RANGE

The domain of polynomial functions, $p(x)$, is all real numbers. The domain of rational functions, $\frac{p(x)}{q(x)}$, is all x except for those values that make $q(x) = 0$. For algebraic functions, we have to exclude only those values for the domain that make the denominator equal to 0 and/or values that result in an even root of a negative number. *Transcendental functions*, or nonalgebraic functions, cannot be expressed in terms of algebraic operations except as an infinite series. Examples are trigonometric functions and exponential functions, such as the function y in $y = 10^x$; however, $y = \sqrt{2}$ is not transcendental because $\sqrt{2}$ is the solution to $y^2 = 2$, an algebraic expression. See Chapter 5 for more on exponential functions. Transcendental functions will have their own limitations on domain values.

The domain of a function can be found through its equation, through its graph, or from its table of values, if the function is defined by a table.

Example

Given the graph of the function $y = x^4 - 4x^3 + \frac{17}{4}x^2 - x - 5$, find the domain and range of the function from the graph shown in Figure 3.1.

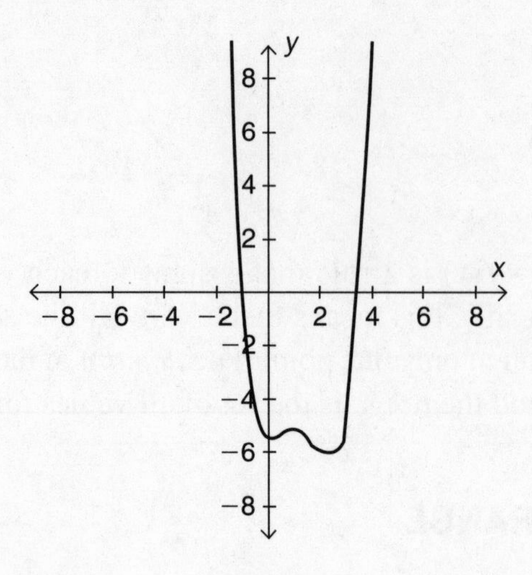

Figure 3.1

Solution

Since this is a polynomial function, the domain is the set of all real numbers. From the graph we see that the range is all y such that $y \geq -6$.

Example

Consider the graph of $y = x^4 - 4x^3 + \frac{17}{4}x^2 - x + a$, where a is a constant. For $a = \{-4, -3, -2, -1, 0, 1, 2, 3, 4\}$, graph the function for these various values for a and see if you can determine a rule for the range of the function.

Solution

Note that this is the same shape as the graph in the preceding example with the value of y changing according to the value of a.

Values of *a*	Range from Graph
-4	$y \geq -5$
-3	$y \geq -4$
-2	$y \geq -3$
-1	$y \geq -2$
0	$y \geq -1$
1	$y \geq 0$
2	$y \geq 1$
3	$y \geq 2$
4	$y \geq 3$

Table 3.1

From Table 3.1, it appears that the range will always be $y \geq (a - 1)$.

Example

Find the domain and range of the function $f(x) = \sqrt{4 - x^2}$.

Solution

Domain: All values of x for which $4 - x^2 \geq 0$, or $-2 \leq x \leq 2$. Range: Since the radicand is nonnegative, the y-value is limited by the domain of $-2 \leq x \leq 2$, so the range must be $0 \leq y \leq 2$.

Example

Find the range and domain of the function $f(x)=\frac{1}{\sqrt{x-2}}$.

Solution

Domain: All values for which $x - 2 > 0$, or $x > 2$. Range: Since the fraction (i.e., the function) cannot equal 0 and the radicand is positive, the range is all positive values, or $y > 0$.

Example

What is the domain of $f(x)=\frac{x-2}{x^2+3x-10}$?

Solution

We first factor the denominator.

$$f(x)=\frac{x-2}{(x-2)(x+5)}$$

Although this simplifies to $f(x)=\frac{1}{x+5}$, we still must exclude $x = 2$ along with excluding $x = -5$. Therefore, the domain is all real values of x, except $x = 2$ and $x = -5$. Graphically, this indicates that $x = -5$ will be a vertical asymptote. There will be a hole in the graph at $x = 2$.

Example

What is the range of $f(x)=(x-a)^{\frac{1}{2}}+b$, where a and b are constants?

Solution

Since the minimum value of $(x-a)^{\frac{1}{2}}$ is zero, then the range must start at b and increase. Hence, the range must be $y \geq b$.

EVEN AND ODD FUNCTIONS

An *even* function is one in which $f(x) = f(-x)$ for all x in the domain of the function. Since (x, y) matches with $(-x, y)$, the graph of the function can be reflected about the y-axis.

An *odd* function is one in which $f(-x) = -f(x)$ for all x in the domain of the function. This reflects a function about the origin.

Knowing whether a function is even, odd, or neither can help in sketching the graph of the function.

INCREASING AND DECREASING FUNCTIONS

If $f(x_1) > f(x_2)$ whenever $x_1 > x_2$, then the function f is increasing. If $f(x_1) < f(x_2)$ whenever $x_1 > x_2$, then the function f is decreasing. A constant function is neither increasing nor decreasing.

EVALUATING FUNCTIONS

Example

Given $f(x) = 3x^2 + 2x - 1$, find (a) $f(1)$ and (b) $f(a)$.

Solution

(a) $f(1) = 3(1)^2 + 2(1) - 1 = 4$

(b) $f(a) = 3(a)^2 + 2(a) - 1 = 3a^2 + 2a - 1$

Example

Given $g(x) = \frac{1}{x}$, find $\frac{g(x+h) - g(x)}{h}$ and simplify.

Solution

$$\frac{g(x+h) - g(x)}{h} = \frac{\frac{1}{x+h} - \frac{1}{x}}{h}$$

We multiply both the numerator and the denominator by $x(x + h)$ to clear out the fractions in the numerator.

$$\left(\frac{x(x+h)}{x(x+h)}\right)\left(\frac{\frac{1}{x+h}-\frac{1}{x}}{h}\right)$$

We now simplify.

$$\frac{x-(x+h)}{h(x(x+h))}=\frac{x-x-h}{hx(x+h)}=\frac{-h}{hx(x+h)}=\frac{-1}{x(x+h)}$$

Example

Given $f(x)=\dfrac{3x-1}{x+2}-1$, find the value of a so that $f(a) = 1$.

Solution

$$f(a)=\frac{3a-1}{a+2}-1=1$$

$$\frac{3a-1}{a+2}=2$$

$$3a-1=2a+4$$

$$a=5$$

COMPOSITE FUNCTIONS

Another operation that can be performed on functions is composition. A *composite function* is made up of multiple functions acting on each other in a specified order. The notation generally used is $f \circ g$, which means $f(g(x))$, and $g \circ f$, which is $g(f(x))$.

Example

If $f(x) = 2x - 3$ and $g(x) = x^2 - 2x + 5$, find (a) $f \circ g$; (b) $g \circ f$; (c) $f(g(2))$; (d) $g(f(-1))$.

Solution

(a) $f \circ g = f(g(x)) = f(x^2 - 2x + 5) = 2(x^2 - 2x + 5) - 3 = 2x^2 - 4x + 7$

(b) $g \circ f = g(f(x)) = g(2x - 3) = (2x - 3)^2 - 2(2x - 3) + 5 = 4x^2 - 16x + 20$

(c) $f(g(2)) = 2(2)^2 - 4(2) + 7 = 7$ *or* evaluate $g(2) = 5$, and then $f(5) = 10 - 3 = 7$

(d) $g(f(-1)) = 4(-1)^2 - 16(-1) + 20 = 40$ *or* evaluate $f(-1) = -5$, and then $g(-5) = 25 + 10 + 5 = 40$

Example

In Table 3.2, what is $g(f(1))$?

x	$f(x)$	$g(x)$
0	0	3
1	2	2
2	4	1
3	5	0

Table 3.2

Solution

From Table 3.2, we see that $f(1) = 2$. We now find $g(2)$ from the table. Therefore, $g(f(1)) = g(2) = 1$.

Example

Is composition of functions commutative? That is, $f(g(x)) = g(f(x))$?

Solution

To check, let's use the functions defined in Table 3.2. Is $f(g(1)) = g(f(1))$? From the previous example, we know $g(f(1)) = 1$. We now find $f(g(1)) = f(2) = 4$. So, $f(g(x)) \neq g(f(x))$ for the functions in Table 3.2. In general, composition of functions is *not* commutative.

SKETCHING PARENT FUNCTIONS

There are some key algebraic functions in mathematics, called *parent functions*, that students should recognize and be able to sketch. Some of these parent functions are $f(x) = x^2$, $f(x) = x^3$, $f(x) = \sqrt{x}$, $f(x) = \frac{1}{x}$, and $f(x) = |x|$, shown in Figures 3.2 through 3.6, respectively. Later chapters discuss some of these functions in greater detail.

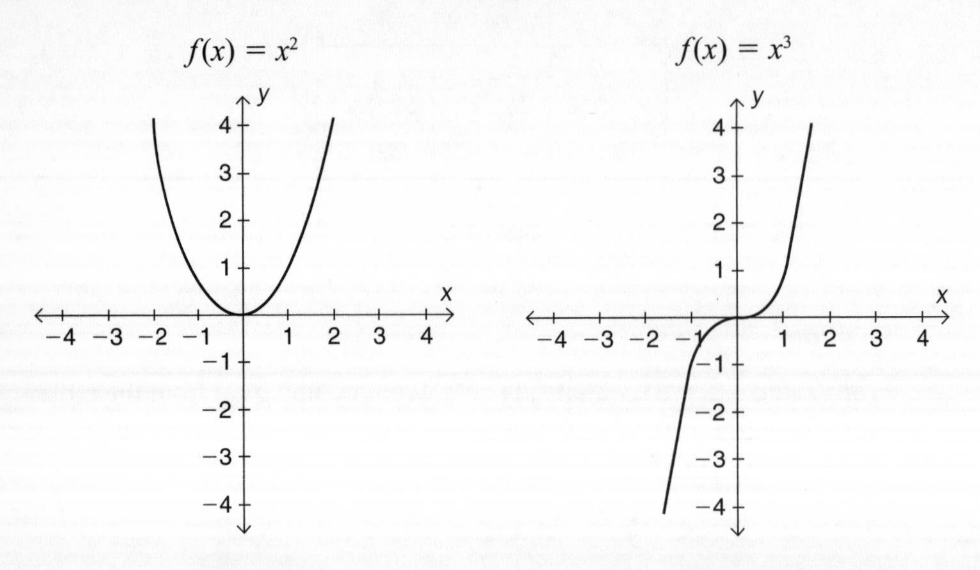

Figure 3.2

Figure 3.3

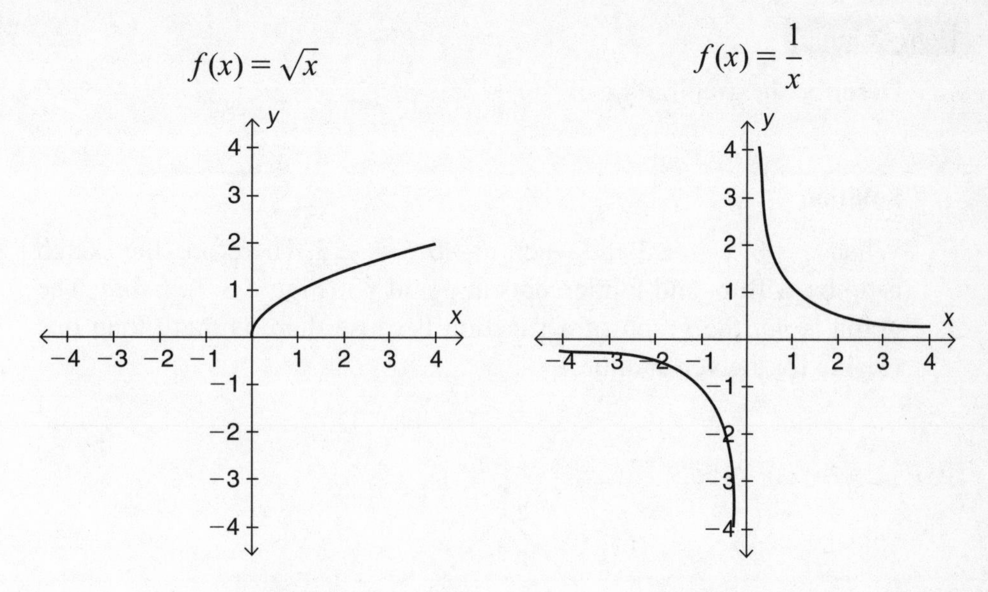

Figure 3.4 **Figure 3.5**

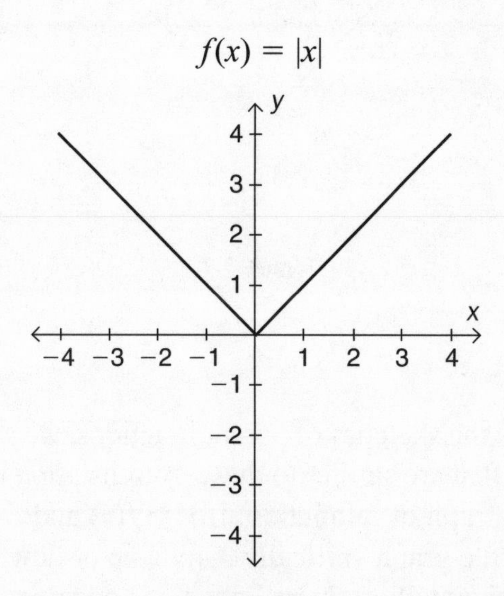

Figure 3.6

Example

Describe the graph of $|x| + |y| = 3$.

Solution

When $x = 0$, $y = \pm 3$ and when $y = 0$, $x = \pm 3$. Therefore, the sketch is a box with x- and y-intercepts at 3 and -3 (Figure 3.7). (*Note*: The graph is *not* the graph of a function because there is more than one y value for a given x value.)

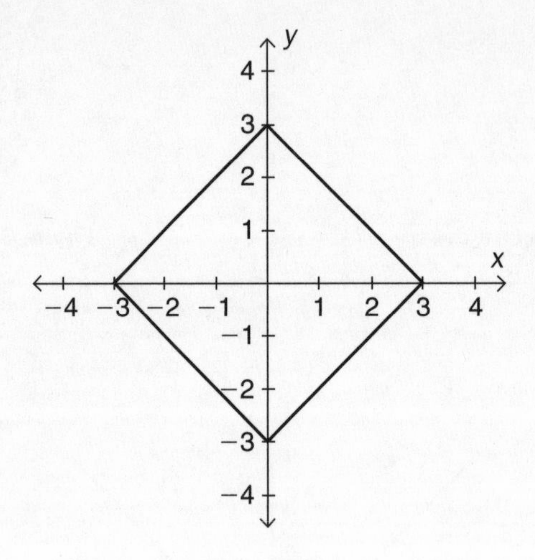

Figure 3.7

TRANSLATIONS

Once we know the sketches of parent functions, we can use translations to sketch functions that are similar to these. A translation is a vertical or horizontal shift of the graph of a function. If $y = f(x)$ and c is a constant, then $y = f(x) + c$ shifts the graph vertically c units up or down and $y = f(x + c)$ shifts the graph horizontally right or left. If c is positive, then the shift is to the left for the vertical shift and up for the horizontal shift. If c is negative, then the shift is to the right for vertical and down for horizontal shifts.

Example

Sketch $f(x) = x^2 + 2$.

Solution

This translates the parent function $f(x) = x^2$ up two units, as shown in Figure 3.8.

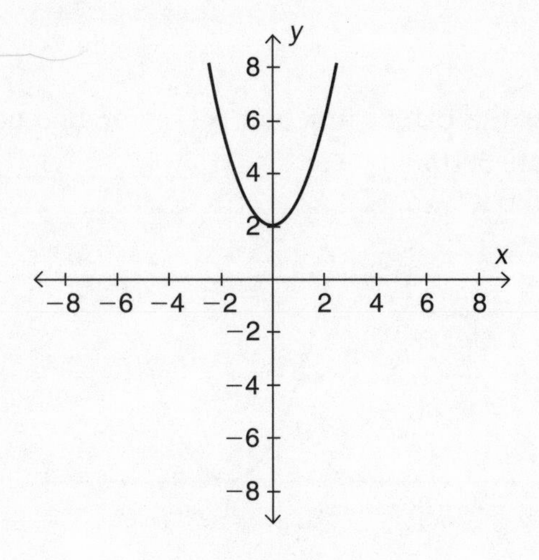

Figure 3.8

Example

Sketch $f(x) = x^2 - 1$.

Solution

This translates the parent function $f(x) = x^2$ down one unit, as shown in Figure 3.9.

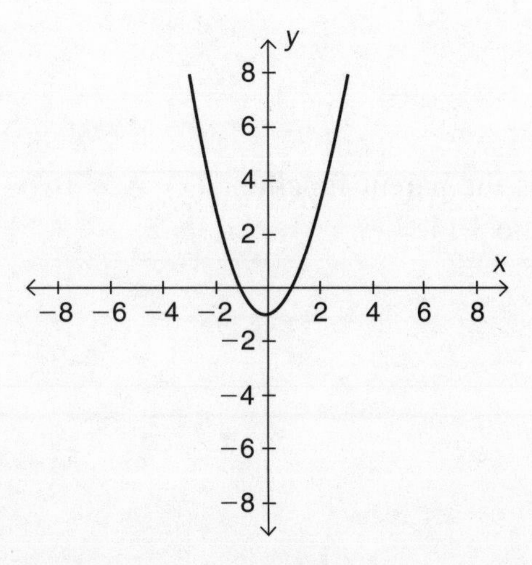

Figure 3.9

Example

Sketch $f(x) = (x - 1)^2$.

Solution

This translates the parent function $f(x) = x^2$ one unit to the right, as shown in Figure 3.10.

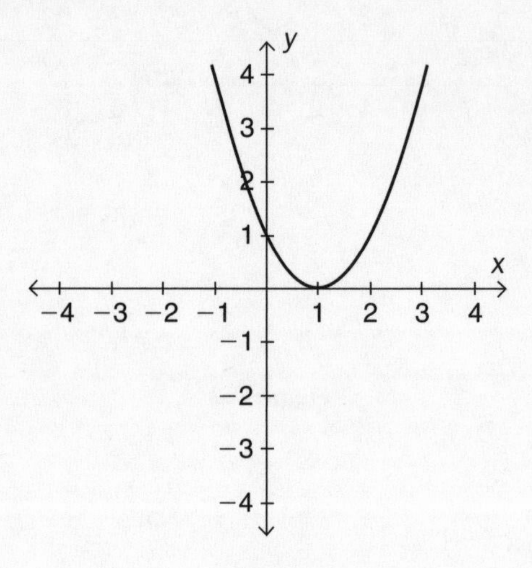

Figure 3.10

Example

Sketch $f(x) = (x + 2)^2$.

Solution

This translates the parent function $f(x) = x^2$ two units to the left, as shown in Figure 3.11.

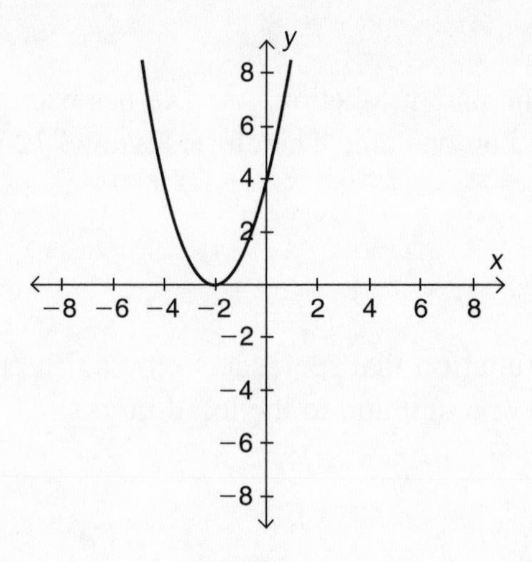

Figure 3.11

To summarize, if $y = f(x)$ is a parent function, then with $c > 0$:

$y = f(x - c)$ translates the function c units to the right.

$y = f(x + c)$ translates the function c units to the left.

$y = f(x) + c$ translates the function c units up.

$y = f(x) - c$ translates the function c units down.

Example

Figure 3.12 is a sketch of which function?

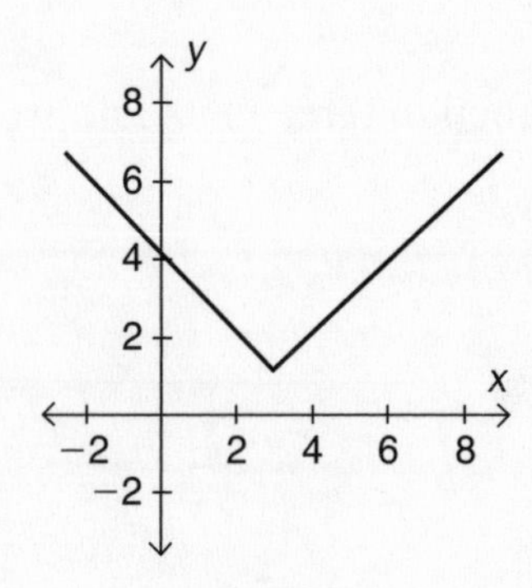

Figure 3.12

Solution

We see that the parent function, $y = |x|$, has been shifted to the right three units and up one unit. Therefore, Figure 3.12 is the graph of $y = |x - 3| + 1$.

Example

What is the function that represents moving the graph of $y = x^2 - 2x + 4$ down one unit and to the left 3 units?

Solution

We first put this equation into a form from which we can recognize how it has been moved from the parent function, $y = x^2$. To do so, we complete the square on the x variable. (See Chapter 2, Algebra Review, for more information on completing the square.)

$y = (x^2 - 2x + \underline{\quad}) + 4 - \underline{\quad}$

$y = (x^2 - 2x + \underline{\ 1\ }) + 4 - \underline{\ 1\ }$

$y = (x - 1)^2 + 3$

To go down one unit, we subtract 1 from the constant term. This changes the function to:

$y = (x - 1)^2 + 3 - 1$

$y = (x - 1)^2 + 2$

To move this new function three units left, we add 3 to the x-term.

$y = (x - 1 + 3)^2 + 2$

$y = (x + 2)^2 + 2$

or, expanding this out,

$y = x^2 + 4x + 6$

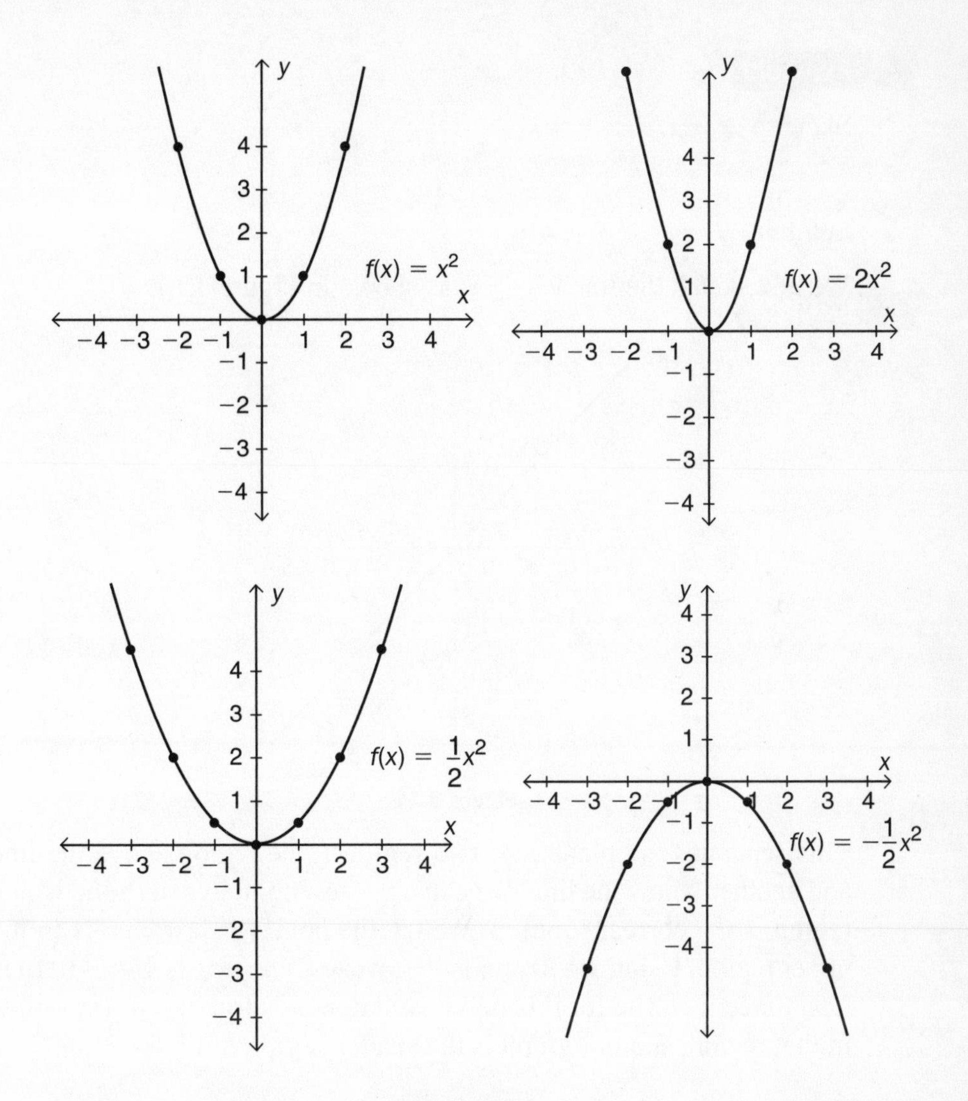

Figure 3.13

Note: If the variable is *multiplied* by a constant a, the resulting graph $y = f(ax)$ is either narrower (if $a > 1$) or wider (if $0 < a < 1$) than the graph of $y = f(x)$. If $a < 0$, the direction of the graph of the function will change from "up/down" to "down/up." Figure 3.13 illustrates the result of multiplying the variable x by a.

GRAPHING INEQUALITIES

Although inequalities are not functions, we can use what we know about sketching functions to sketch the graph of most inequalities.

Example

Sketch $y \geq -x$.

Solution

We first sketch the line $y = -x$, as shown in Figure 3.14.

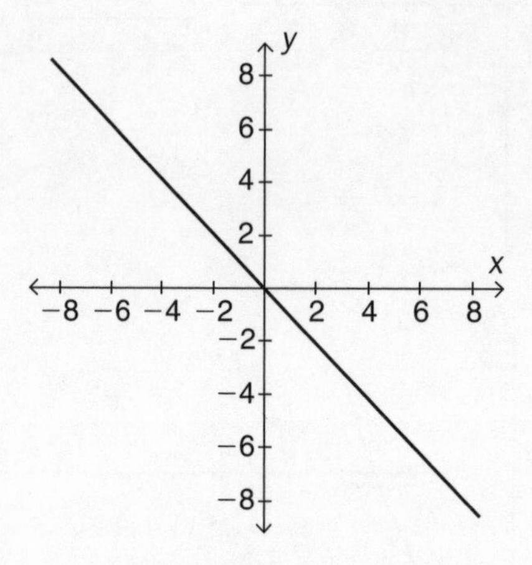

Figure 3.14

This separates the plane into two regions: one region above the line and another below the line. We can use a test point to determine which region is the correct solution. We try the point (0, 1) which is in the upper region. Using $x = 0$ and $y = 1$, we test $y \geq -x$. Is $1 \geq -0$ true? Yes. Since it is true for this point, we know it is true for all the points in that region, and our graph is the shaded region of Figure 3.15.

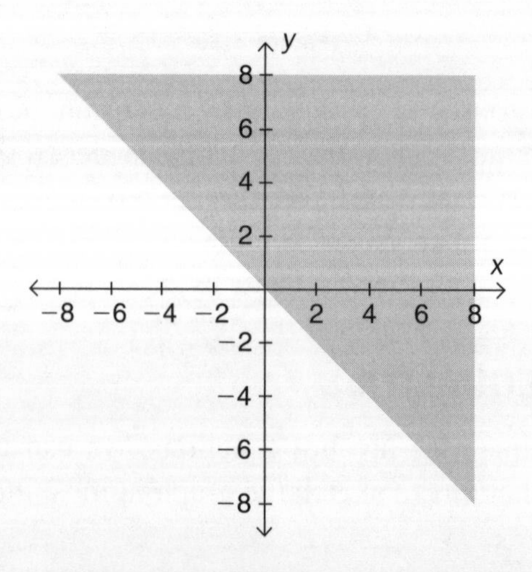

Figure 3.15

Example

Sketch $y \leq -x^2 + 2x + 3$.

Solution

We first sketch $y = -x^2 + 2x + 3$ by completing the square and using this information to sketch from the parent function $y = x^2$. (See Chapter 2, Algebra Review, for more information on completing the square.)

$y = -(x^2 - 2x + \underline{\quad}) + 3 + \underline{\quad}$

$y = -(x^2 - 2x + \underline{\ 1\ }) + 3 + \underline{\ 1\ }$

$y = -(x - 1)^2 + 4$

This moves the parent function to the right one unit and up four units. The graph opens downward since the coefficient of x^2 is negative. The graph crosses the y-axis when $x = 0$ or at $y = 3$. The graph crosses the x-axis when $y = 0$, or when

$-x^2 + 2x + 3 = 0$

Multiplying through by -1, factoring, and solving, we get

$x^2 - 2x - 3 = 0$

$(x + 1)(x - 3) = 0$

$x = -1 \quad x = 3$

We now can sketch the graph (Figure 3.16) using the information from the translation as well as the x and y intercepts.

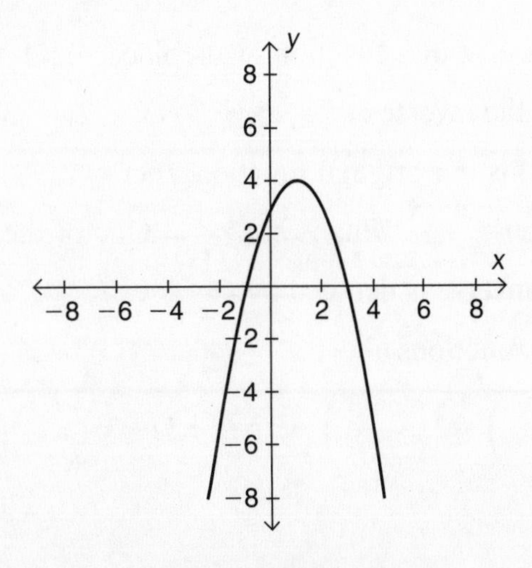

Figure 3.16

The graph divides the plane into two regions. We want to know where $y \leq -x^2 + 2x + 3$. We use a test point in one of the regions, say (0, 0). We substitute $x = 0$ and $y = 0$ into the inequality to see whether it is true or false. Is $0 \leq -(0)^2 + 2(0) + 3$ true? Yes. Since it is true for (0, 0), then the region that contains (0, 0) is the region that contains all points for the inequality, as shown in Figure 3.17

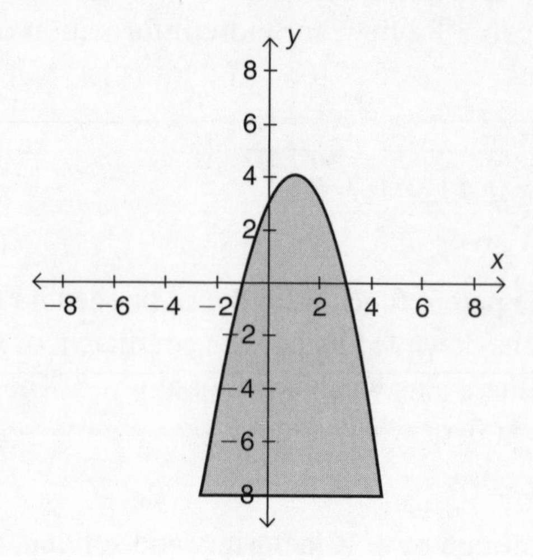

Figure 3.17

Note: In the last two examples, the equality was included. If the inequality is strict, the boundary line would be a dashed line.

INVERSE FUNCTIONS

To form the inverse of a function, we replace x and y and solve for the "new" y. Therefore, the inverse of $y = 2x - 3$ is $x = 2y - 3$. We solve for y so that $y = \frac{x+3}{2}$. If $f(x)$ is our original function, $f(x) = 2x - 3$, then its inverse is denoted by $f^{-1}(x) = \frac{x+3}{2}$. This is *not* $\frac{1}{f(x)}$! One of the characteristics of functions and their inverse is denoted by $f \circ f^{-1} = f^{-1} \circ f = x$. We can verify this relation for the functions above.

$$f \circ f^{-1} = f\left(\frac{x+3}{2}\right) = 2\left(\frac{x+3}{2}\right) - 3 = (x+3) - 3 = x$$

$$f^{-1} \circ f = f^{-1}(2x-3) = \frac{(2x-3)+3}{2} = \frac{2x}{2} = x$$

How do we know whether a function has an inverse that is a function? Since vertical lines are a test that can be used to determine if a graph describes a function, then we can use an analogous *horizontal line test* on the original function to determine whether the function has an inverse that is a function. Functions of this type are called one-to-one (or 1–1) functions since there is not only one y for every x, there is also only one x for every y. Sometimes we restrict the domain of the function so that its inverse is a function (or so that it passes the horizontal line test).

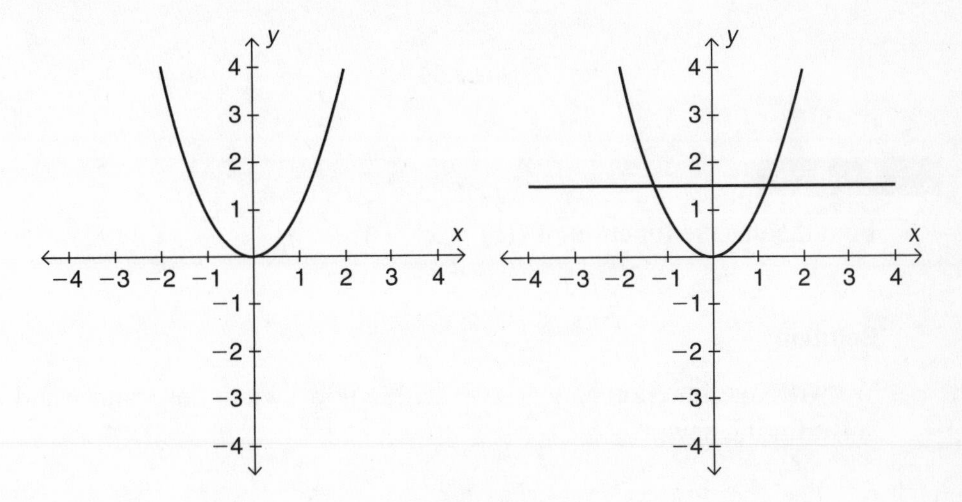

Figure 3.18

We see from the graph of $y = x^2$ in Figure 3.18 that it is a function, but its inverse is not a function since $y = x^2$ fails the horizontal line test. We can also see this from the inverse equation: $x = y^2$ or $y = \pm\sqrt{x}$, which is not a function since one value for x will give two values for y. But, if we restrict the domain of the original function, $y = x^2$, to $x \geq 0$, then the inverse will be a function, as seen in Figure 3.19.

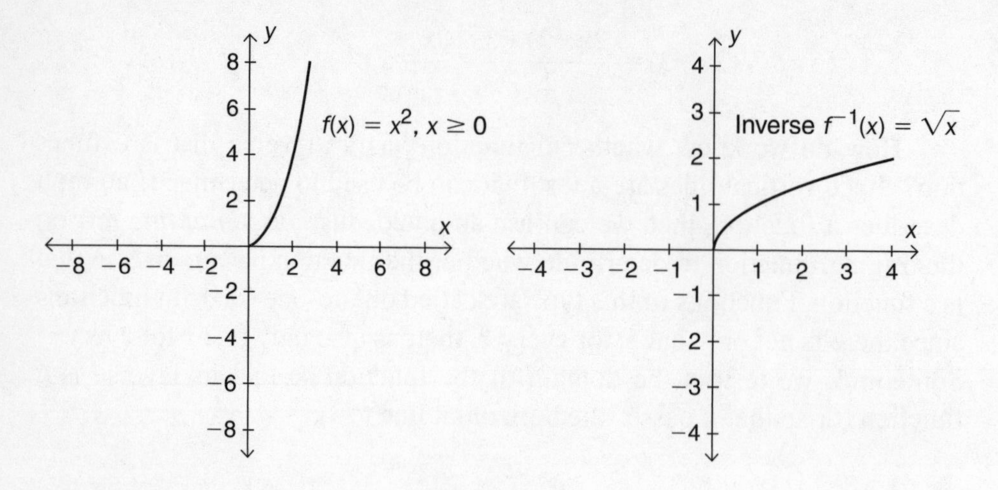

Figure 3.19

Example

Find the inverse function of $f(x) = x^3 - 1$.

Solution

We write the function as $y = x^3 - 1$. We now switch the x and y and solve for the new y.

$x = y^3 - 1$

$y^3 = x + 1$

$y = \sqrt[3]{x+1}$

or $\quad f^{-1}(x) = \sqrt[3]{x+1}$

We can verify that these are inverses by checking whether $f \circ f^{-1} = f^{-1} \circ f = x$.

$$f \circ f^{-1} = f\left(\sqrt[3]{x+1}\right) = \left(\sqrt[3]{x+1}\right)^3 - 1 = x + 1 - 1 = x$$

$$f^{-1} \circ f = f^{-1}(x^3 - 1) = \sqrt[3]{(x^3-1)+1} = \sqrt[3]{x^3} = x$$

Example

If $f(x) = \sqrt[4]{2x-5}$, find $f^{-1}(-4)$.

Solution

We first must find $f^{-1}(x)$. We write the function as $y = \sqrt[4]{2x - 5}$, switch x and y, and solve for the new y.

$$x = \sqrt[4]{2y - 5}$$

$$x^4 = 2y - 5$$

$$y = \frac{x^4 + 5}{2}$$

$$\text{or} \quad f^{-1}(x) = \frac{x^4 + 5}{2}$$

We now evaluate $f^{-1}(-4)$.

$$f^{-1}(-4) = \frac{(-4)^4 + 5}{2} = \frac{256 + 5}{2} = \frac{261}{2}$$

PIECEWISE FUNCTIONS

A piecewise function is a function that uses different expressions for different parts of its domain. For example, f is a piecewise function.

$$f(x) = \begin{cases} 2x - 3, & x \geq 2 \\ 2, & 0 < x < 2 \\ -x^3, & x \leq 0 \end{cases}$$

Example

Find the range of the piecewise function defined above.

Solution

For $x \geq 2$, the y values are all greater than or equal to 1, since the function is equal to $2x - 3$. For x values between 0 and 2, the y value is equal to the constant 2. If $x \leq 0$, the function is defined as $-x^3$, which is nonnegative, since $x \leq 0$. Therefore, there are nonnegative values for y. The range is $y \geq 0$.

CHAPTER 4
Polynomial and Rational Functions

Chapter 4
Polynomial and Rational Functions

DIVISION OF POLYNOMIALS

In any rational function, if the degree of the numerator is greater than or equal to the degree of the denominator, then we can divide using a process similar to the long division algorithm from arithmetic. For example, to divide $\frac{x^4 - 4x^2 + 7x - 15}{x + 4}$, we write $x + 4\overline{)x^4 + 0x^3 - 4x^2 + 7x - 15}$, filling in any missing terms with a coefficient of 0.

We divide the first term of the divisor (x) into the first term of the dividend (x^4).

$$\begin{array}{r} x^3 \\ x + 4\overline{)x^4 + 0x^3 - 4x^2 + 7x - 15} \end{array}$$

We continue with the division algorithm and multiply, subtract, and bring down the next term.

$$\begin{array}{l} x^3 \\ x + 4\overline{)x^4 + 0x^3 - 4x^2 + 7x - 15} \\ \underline{x^4 + 4x^3} \\ -4x^3 - 4x^2 \end{array}$$

We start the algorithm again and divide the first term of the divisor (x) into the first term of this subtracted expression ($-4x^3$). We then multiply, subtract, and bring down the next term.

$$\begin{array}{l} x^3 - 4x^2 \\ x + 4\overline{)x^4 + 0x^3 - 4x^2 + 7x - 15} \\ \underline{x^4 + 4x^3} \\ -4x^3 - 4x^2 \\ \underline{-4x^3 - 16x^2} \\ +12x^2 + 7x \end{array}$$

Continuing this algorithm until we can no longer divide, we obtain

$$
\begin{array}{r}
x^3 - 4x^2 + 12x - 41 \quad \text{R. } 149 \\
x+4\overline{)x^4 + 0x^3 - 4x^2 + 7x - 15} \\
\underline{x^4 + 4x^3} \qquad\qquad\qquad\qquad \\
-4x^3 - 4x^2 \qquad\qquad\quad \\
\underline{-4x^3 - 16x^2} \qquad\qquad\quad \\
+12x^2 + 7x \qquad\quad \\
\underline{+12x^2 + 48x} \qquad\quad \\
-41x - 15 \\
\underline{-41x - 164} \\
+149
\end{array}
$$

The remainder is usually expressed as a fraction of the divisor. Here it would be $\frac{149}{x+4}$.

SYNTHETIC DIVISION

Another process that can be used in division of polynomials is synthetic division. In this process we use only the coefficients and proceed through the algorithm. Since the algorithm uses subtraction, and subtraction is a process of changing the sign of the subtrahend and combining, we can change the sign of 4 in the divisor of $x + 4$, divide synthetically by -4, and then combine the terms. We write the coefficients of the variables in the dividend, putting in 0 for coefficients of missing terms. For the above example, we have

$$
-4 \,|\quad 1 \quad 0 \quad -4 \quad 7 \quad -15
$$

We bring down the first coefficient (which is 1), and then continue with the process of multiplying by the divisor (which is -4), placing the product in the next column, combining, and continue with multiplying, placing the product in the next column, and combining.

$$
\begin{array}{r|rrrrr}
-4 & 1 & 0 & -4 & 7 & -15 \\
 & & -4 & 16 & -48 & +164 \\
\hline
 & \mathbf{1} & \mathbf{-4} & \mathbf{12} & \mathbf{-41} & +149
\end{array}
$$

The last entry is always the remainder and the other numbers are the coefficients of the variables in the quotient polynomial, beginning with one degree less than the numerator polynomial. From the synthetic division above, we have the polynomial and remainder.

$$\mathbf{1}x^3 - \mathbf{4}x^2 + \mathbf{12}x - \mathbf{41} + \frac{149}{x+4}$$

REMAINDER THEOREM

> If a polynomial $p(x)$ is divided by the binomial $x - c$, the remainder is $p(c)$.

Recall that the remainder from division is the last term of synthetic division. In our previous example, $\frac{x^4 - 4x^2 + 7x - 15}{x+4}$, the remainder was 149. To use the Remainder Theorem, we take the polynomial in the numerator, $p(x) = x^4 - 4x^2 + 7x - 15$, and evaluate it at $x = -4$.

$$p(-4) = (-4)^4 - 4(-4)^2 + 7(-4) - 15 = 149$$

This matches the remainder we had from synthetic division.

In a similar fashion, dividing $\frac{2x^4 - 3x^3 + 2x - 1}{x-3}$ through synthetic division, we have

$$\begin{array}{r|rrrrr} 3 & 2 & -3 & 0 & 2 & -1 \\ & & 6 & 9 & 27 & 87 \\ \hline & \mathbf{2} & \mathbf{3} & \mathbf{9} & \mathbf{29} & 86 \end{array}$$

The remainder is 86. We can use the Remainder Theorem for $\frac{2x^4 - 3x^3 + 2x - 1}{x-3}$ by evaluating the numerator polynomial, $p(x) = 2x^4 - 3x^3 + 2x - 1$, at $x = 3$.

$$p(3) = 2(3)^4 - 3(3)^3 + 2(3) - 1 = 86$$

Example

Find the remainder when $p(x) = x^3 - 2x^2 + x - 5$ is divided by $x + 1$.

Solution

We can solve this problem two ways: use the Remainder Theorem or divide synthetically. If we use the Remainder Theorem, we evaluate $p(-1)$ to get

$p(-1) = (-1)^3 - 2(-1)^2 + (-1) - 5 = -9$

Therefore, the remainder is -9. If we divide synthetically, we have

$$\begin{array}{r|rrrr} -1 & 1 & -2 & 1 & -5 \\ & & -1 & 3 & -4 \\ \hline & \mathbf{1} & \mathbf{-3} & \mathbf{4} & \mathbf{-9} \end{array}$$

The last term, -9, is the remainder.

FACTOR THEOREM

> If the remainder is zero when $f(x)$ is divided by $x - c$, then $x - c$ is a factor of $f(x)$.

Example

Is $x - 3$ a factor of $x^3 - x^2 - 4x - 6$?

Solution

We can use the Factor Theorem and the Remainder Theorem to answer this question. If $f(x) = x^3 - x^2 - 4x - 6$, we evaluate $f(3) = (3)^3 - (3)^2 - 4(3) - 6 = 0$. Since the remainder is zero, then we use the Factor Theorem to determine that $x - 3$ is a factor of $x^3 - x^2 - 4x - 6$. We can also use synthetic division to see if the remainder is zero when the polynomial is divided by $x - 3$.

$$\begin{array}{r|rrrr} 3 & 1 & -1 & -4 & -6 \\ & & 3 & 6 & 6 \\ \hline & \mathbf{1} & \mathbf{2} & \mathbf{2} & \mathbf{0} \end{array}$$

Since the remainder is zero, then $x - 3$ is a factor of $x^3 - x^2 - 4x - 6$.

We can use this information to find the remaining zeros (or roots) of the polynomial. From the synthetic division process, we see that the remaining polynomial is $x^2 + 2x + 2$, which we can solve by the quadratic formula.

$$x = \frac{-2 \pm \sqrt{(2)^2 - 4(1)(2)}}{2(1)} = \frac{-2 \pm \sqrt{-4}}{2} = -1 \pm i$$

Therefore, the zeros of $x^3 - x^2 - 4x - 6$ are $3, -1 + i, -1 - i$.

Example

Is $(x + 2)$ a factor of $(4x^3 + 11x^2 + 21x + 22)$?

Solution

We can answer this question with the Remainder Theorem by evaluating $f(-2)$ for $f(x) = 4x^3 + 11x^2 + 21x + 22$ and determining the remainder.

$f(-2) = 4(-2)^3 + 11(-2)^2 + 21(-2) + 22$

$f(-2) = -8$

Therefore, $(x + 2)$ is *not* a factor of $(4x^3 + 11x^2 + 21x + 22)$. We could also have divided synthetically by (-2) and checked to see whether the remainder is zero.

RATIONAL ROOT THEOREM

Given $f(x) = a_n x^n + a_{n-1}x^{n-1} + \ldots + a_1 x + a_0$. If $\frac{p}{q}$ is a root (zero) of $f(x)$, then p is a factor of a_0 and q is a factor of a_n.

This useful theorem from algebra allows us to check for rational roots (or zeros) of a polynomial.

Example

Find all possible rational roots of $f(x) = 5x^4 - 3x^3 + 6x^2 - x + 18$.

Solution

From the Rational Root Theorem, we know that any rational roots to a polynomial must be of the form $\pm\frac{p}{q}$, where p is a factor of a_0 (the constant) and q is a factor of a_n (the coefficient of the highest power). In this problem, $p = 18$ and $q = 5$. We now list the factors of p and q.

Factors of 18: $\pm1, \pm2, \pm3, \pm6, \pm9, \pm18$

Factors of 5: $\pm1, \pm5$

Therefore, any rational roots to the polynomial function $f(x) = 5x^4 - 3x^3 + 6x^2 - x + 18$ will be:

$$\pm\frac{1}{1}, \pm\frac{1}{5}, \pm\frac{2}{1}, \pm\frac{2}{5}, \pm\frac{3}{1}, \pm\frac{3}{5}, \pm\frac{6}{1}, \pm\frac{6}{5}, \pm\frac{9}{1}, \pm\frac{9}{5}, \pm\frac{18}{1}, \pm\frac{18}{5}$$

Note that the Rational Root Theorem gives us all the *possible* rational roots, but not which of them is an actual root. Here there are 24 possible rational roots, but since the polynomial is only a fourth-order polynomial, there are only four roots, not all of which will be necessarily rational.

ZEROS OF POLYNOMIALS

A zero of a polynomial function is the x-value where the graph of the function either crosses or touches the x-axis. For example, in Figure 4.1, there are three zeros of the polynomial, but one of the zeros (at $x = 1$) has multiplicity two.

The multiplicity of a zero is the number of times the value is a zero of the function. Graphically, a multiplicity of two (or any even positive integer) indicates that the graph of the function just touches the x-axis, but does not cross it.

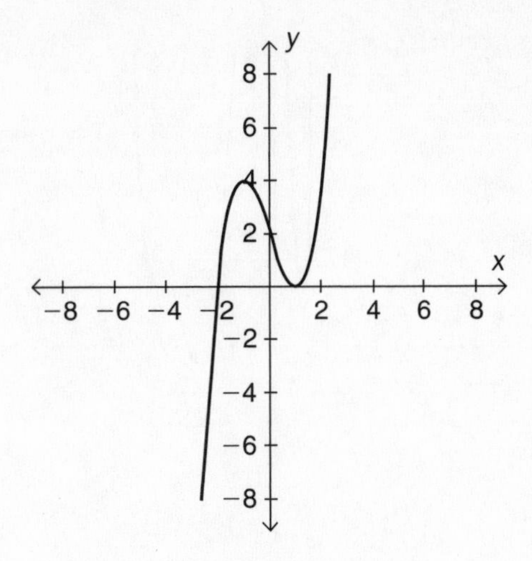

Figure 4.1

Example

Find the zeros and multiplicities for

$f(x) = (2x + 1)^2 (x - 2)^3 (x - 1)$

and sketch the function.

Solution

The zeros are:

$x = -\frac{1}{2}$ with multiplicity two

$x = 2$ with multiplicity three

$x = 1$ with multiplicity one

The zeros separate the coordinate axes into regions. We can use a test point in each region to see whether the graph is in the upper or lower parts of each region. This gives us an idea of how the graph looks. Figure 4.2 is a sketch of the graph for $y \leq 4$. If we set $x = 0$, we can determine where the graph crosses the y-axis, which is at (0, 8).

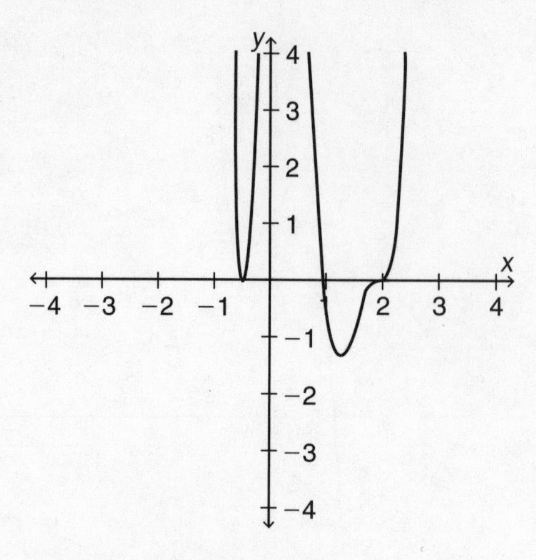

Figure 4.2

PUTTING IT ALL TOGETHER

Example

Find the zeros and their multiplicities for $f(x) = x^4 - 4x^3 - 2x^2 + 12x + 9$.

Solution

To find the zeros and their multiplicities, we first must obtain the factors of the polynomial. We use the Rational Root Theorem to help us find any rational zeros. Possible rational roots are ± 1, ± 3, and ± 9. We can divide synthetically by each or use the Remainder Theorem to determine which, if any, of these are zeros. Using the zeros will help us find the factors. Here we use the Remainder Theorem and the possible rational roots for $f(x) = x^4 - 4x^3 - 2x^2 + 12x + 9$.

$f(1) = 16$	Therefore 1 is not a zero and $(x - 1)$ is not a factor.
$f(-1) = 0$	Therefore -1 is a zero and $(x + 1)$ is a factor.
$f(3) = 0$	Therefore 3 is a zero and $(x - 3)$ is a factor.
$f(-3) = 144$	Therefore -3 is not a zero and $(x + 3)$ is not a factor.
$f(9) = 3600$	Therefore 9 is not a zero and $(x - 9)$ is not a factor.
$f(-9) = 9216$	Therefore -9 is not a zero and $(x + 9)$ is not a factor.

We can now divide synthetically by one of the known zeros (say, 3) to help us get a remaining polynomial.

$$\begin{array}{r|rrrrr} 3 & 1 & -4 & -2 & 12 & 9 \\ & & 3 & -3 & -15 & -9 \\ \hline & \mathbf{1} & \mathbf{-1} & \mathbf{-5} & \mathbf{-3} & \mathbf{0} \end{array}$$

Dividing this result synthetically by the other zero (-1), we get

$$\begin{array}{r|rrrr} -1 & 1 & -1 & -5 & -3 \\ & & -1 & 2 & 3 \\ \hline & \mathbf{1} & \mathbf{-2} & \mathbf{-3} & \mathbf{0} \end{array}$$

The last line represents the coefficients of the remaining polynomial, with degree 2.

$1x^2 - 2x - 3$

We can factor this into $(x - 3)(x + 1)$. We have two factors each of $(x - 3)$ and $(x + 1)$. The given polynomial can be factored into the binomials $(x + 1)^2 (x - 3)^2$. Therefore, $x = -1$ is a zero with multiplicity two and $x = 3$ is a zero with multiplicity two.

SKETCHING POLYNOMIALS

Descartes' Rule of Signs can be used to determine the number of possible positive and negative real solutions to a function. The rule states:

> The number of changes in signs from term to term gives the possible number of positive real roots, minus an even number and the number of sign changes from term to term in $f(-x)$ gives us the possible number of negative real roots, minus an even number.

The reason we may have to subtract an even number is to account for possible complex solutions, which are always in pairs. It is usually best to write the polynomial in descending powers.

Example

Determine the possible number of positive and negative real roots of the following function.

$f(x) = 3x^7 + 2x^6 - x^5 - 4x^4 + 5x^3 - 2x^2 + 6x - 1$

Solution

We notice that there are five sign changes from term to term. Therefore, there will be 5, 3, or 1 positive real roots.

To determine the possible number of negative real roots, we go through the same process, except we examine the change in signs for $f(-x)$.

$f(-x) = 3(-x)^7 + 2(-x)^6 - (-x)^5 - 4(-x)^4 + 5(-x)^3 - 2(-x)^2 + 6(-x) - 1$

or simplified

$f(-x) = -3x^7 + 2x^6 + x^5 - 4x^4 - 5x^3 - 2x^2 - 6x - 1$

There are two sign changes, which means there will be 2 or 0 negative real roots to the function.

RATIONAL FUNCTIONS

A rational function is a function of the type $f(x) = \dfrac{p(x)}{q(x)}$, where $p(x)$ and $q(x)$ are polynomials and $q(x) \neq 0$.

Vertical Asymptotes and Holes

To sketch rational functions, we must pay attention to the values of x that make $q(x) = 0$. These values determine vertical asymptotes for the graph. Graphs of rational functions *never* cross vertical asymptotes, since this would make the denominator polynomial equal to zero. If there is an x value for which $p(x) = q(x) = 0$, then there is a "hole" in the graph at that x value.

Horizontal Asymptotes

Rational functions may also have horizontal asymptotes. These asymptotes let us know if the graph is approaching a horizontal line as the absolute value of x gets larger. Graphs *may* cross horizontal asymptotes. There are three cases to consider for horizontal asymptotes.

Case 1

The degree of the numerator is less than the degree of the denominator. Since the denominator is getting larger faster than the numerator, the fraction (i.e., the rational function) approaches 0 as the absolute value of x gets larger. Hence, the horizontal line $y = 0$ is a horizontal asymptote.

All of the following functions have a horizontal asymptote of $y = 0$.

$$f(x) = \frac{1}{2x+3},\ f(x) = \frac{x^3}{2x^5+3},\ f(x) = \frac{x}{x^2-1}$$

Variation of Case 1: If the function is a variation of the rational function, then we can just use what we know about translations to find the horizontal asymptote. For example, the horizontal asymptotes of the following functions are $y = 5$ and $y = -7$, respectively.

$$f(x) = \frac{1}{2x+3} + 5 \text{ and } f(x) = \frac{x^3}{2x^5+3} - 7$$

The fractional part of the functions goes to zero as $x \to \infty$ or $x \to -\infty$ so that the total functions approach the constants at the end. This is the same as translating the rational function up or down the constant amount.

Case 2

The degree of the numerator equals the degree of the denominator. This tells us that the coefficients of the highest degree from the numerator and denominator determine the asymptote.

$f(x) = \dfrac{x}{2x+3}$ will have a horizontal asymptote of $y = \dfrac{1}{2}$, where 1 and 2 are the coefficients of x from the numerator and denominator, respectively.

$f(x) = \dfrac{2x^3 - 5x + 2}{5x^3 + 6x - 3}$ will have a horizontal asymptote of $y = \dfrac{2}{5}$, where 2 and 5 are the coefficients of x^3 from the numerator and denominator, respectively.

Variation of Case 2: If the function is a variation of the rational function, then we can just use what we know about translations to find the horizontal asymptote. For example, the horizontal asymptote of $f(x) = \dfrac{x}{2x+3} + 5$ is $y = \dfrac{11}{2}\left(\text{or } y = \dfrac{1}{2} + 5\right)$. For $f(x) = \dfrac{2x^4}{5x^4+3} - 7$, the horizontal asymptote is $y = -\dfrac{33}{5}\left(\text{or } y = \dfrac{2}{5} - 7\right)$. The fractional part of the functions follows the rule in Case 2, but the total functions approach the value from Case 2 combined with the constants at the end of the function. This is the same as translating the rational function up or down the constant amount.

Case 3

The degree of the numerator is greater than the degree of the denominator. Since the numerator is getting larger at a much faster rate than the denominator, there is no horizontal asymptote. The y-value (i.e., the rational function) just keeps getting larger and larger as the absolute value of x gets larger.

Oblique Asymptotes

If the degree of the numerator is one degree larger than the degree of the denominator, then there is a nonhorizontal line as an asymptote, called an *oblique* asymptote. We can get the equation of that asymptote by dividing the rational function and ignoring the remainder. For example,

$f(x) = \dfrac{x^2 + 2x - 5}{x - 3}$ will have an oblique asymptote of $y = x + 5$, the quotient without the remainder.

$$
\begin{array}{r}
x + 5 + \dfrac{10}{x-3} \\
x - 3\overline{)x^2 + 2x - 5} \\
\underline{x^2 - 3x} \\
+5x - 5 \\
\underline{+5x - 15} \\
+10
\end{array}
$$

Sketching Rational Functions

We can use asymptotes, along with intercepts and test points, to help us sketch rational functions.

Example

Sketch $f(x) = \dfrac{1}{2x + 3}$.

Solution

Intercepts: $\left(0, \dfrac{1}{3}\right)$

Asymptotes:

Vertical: We set the denominator equal to zero and solve to get the vertical asymptotes.

$2x + 3 = 0$ or $x = -\dfrac{3}{2}$ is a vertical asymptote

Horizontal: We use Case 1. Since the degree of the numerator is less than the degree of the denominator, the line $y = 0$ is a horizontal asymptote.

Figure 4.3 shows the sketch of a graph with these parameters.

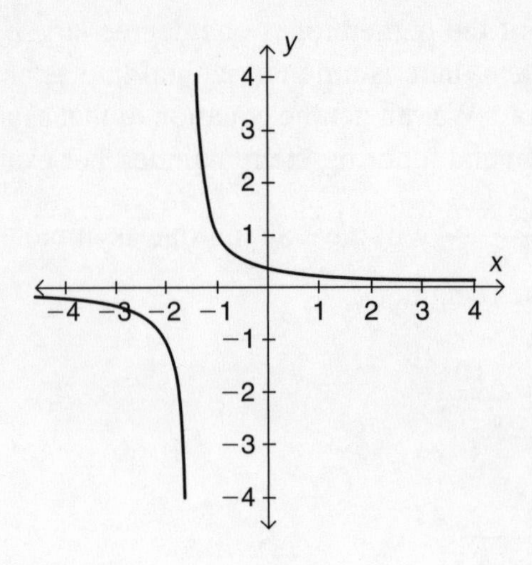

Figure 4.3

Example

Sketch $f(x) = \dfrac{x-2}{x^2-x-2}$.

Solution

Intercepts: (0, 1)

Asymptotes: Since the degree of the numerator is less than the degree of the denominator, the line $y = 0$ is a horizontal asymptote. The denominator equals zero at $x = -1$ and $x = 2$. There is an asymptote at $x = -1$, but at $x = 2$, the numerator is also zero, so $x = 2$ is a hole in the graph. Figure 4.4 shows a sketch of the graph of $f(x) = \dfrac{x-2}{x^2-x-2}$.

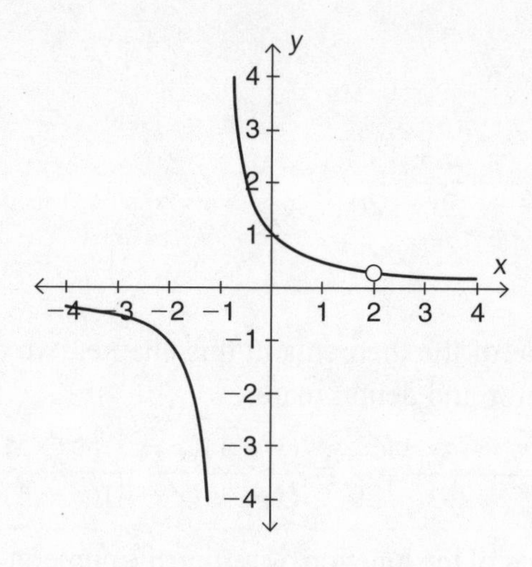

Figure 4.4

Example

Find a possible rational function for the graph shown in Figure 4.5.

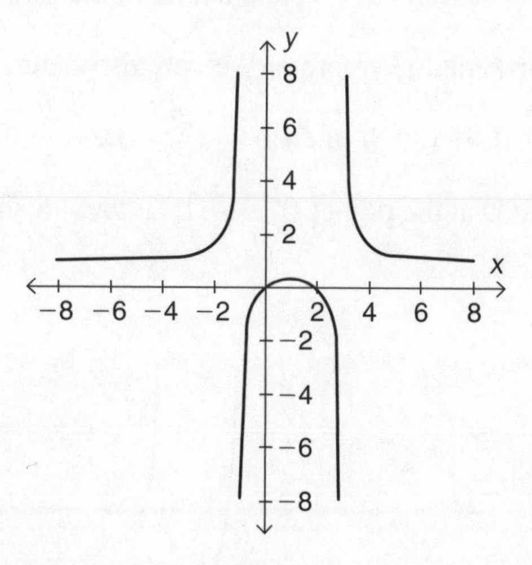

Figure 4.5

Solution

It appears from the graph that $x = -1$ and $x = 3$ are vertical asymptotes, and $y = 1$ is a horizontal asymptote. There are x-intercepts at $x = 0$ and $x = 2$. Therefore, a possible rational function could be:

$$y = \frac{x(x-2)}{(x+1)(x-3)}$$

Example

Sketch

$$y = \frac{x^3 - 7x^2 + 7x + 15}{x^3 - 10x^2 + 29x - 20}$$

Solution

By using some of the theorems in this chapter, we can get the factors of the numerator and denominator.

$$y = \frac{x^3 - 7x^2 + 7x + 15}{x^3 - 10x^2 + 29x - 20} = \frac{(x + 1)(x - 3)(x - 5)}{(x - 1)(x - 4)(x - 5)}$$

We get the zeros of the function by setting the numerator equal to zero and solving. We get the vertical asymptotes by setting the denominator equal to zero and solving. The function has zeros at $x = -1$ and $x = 3$, vertical asymptotes at $x = 1$ and $x = 4$, and a hole in the graph at $x = 5$. The line $y = 1$ is a horizontal asymptote. We can also determine the y-intercept by letting $x = 0$. If $x = 0$, then $y = -\frac{15}{20}$ or $-\frac{3}{4}$. Therefore, the graph crosses the y-axis at the point $\left(0, -\frac{3}{4}\right)$. Figure 4.6 shows a sketch of this function.

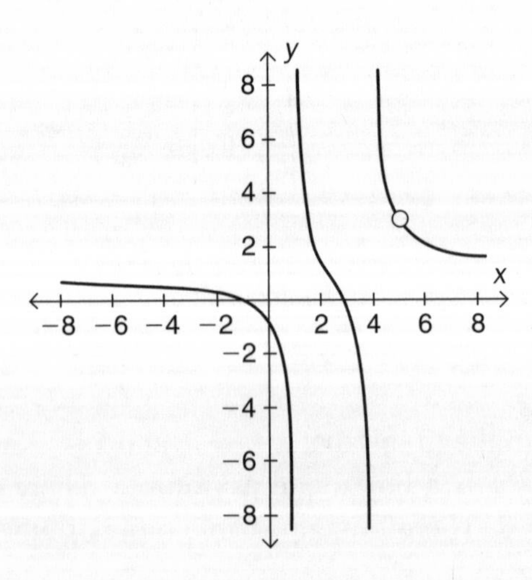

Figure 4.6

INTERSECTION OF GRAPHS

If f and g are two graphs, then their points of intersection occur where $f(x) - g(x) = 0$.

Example

Given the graphs of f and g shown in Figure 4.7 as a line and parabola, respectively, with points of intersection at $x = a$ and $x = b$. If h is defined as $h(x) = f(x) - g(x)$, where is (a) $h(x) > 0$, (b) $h(x) = 0$, and (c) $h(x) < 0$?

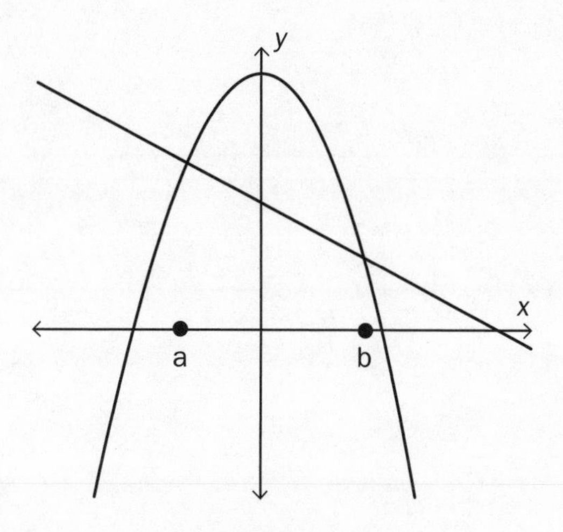

Figure 4.7

Solution

(a) $h(x) > 0$ is where $f(x) - g(x) > 0$ or where $f(x) > g(x)$, where the line is above the parabola. From the graph, we see that f is greater than g for $(-\infty, a)$ and (b, ∞).

(b) $h(x) = 0$ is where $f(x) - g(x) = 0$ or where $f(x) = g(x)$. This occurs at the points of intersection of the line and the parabola at $x = a$ and at $x = b$.

(c) $h(x) < 0$ is where $f(x) - g(x) < 0$ or where $f(x) < g(x)$, where the line is below the parabola. From the graph, we see that f is less than g on the interval $x = (a, b)$.

CHAPTER 5
Trigonometry

Chapter 5
Trigonometry

Given a unit circle in a rectangular coordinate system with the center at the origin, then for any point $P(x, y)$ on the circle, we can define the ratios listed below. We use θ as the central angle, with the x-axis as one ray and the ray through the center and the point P as the other ray. See Figure 5.1.

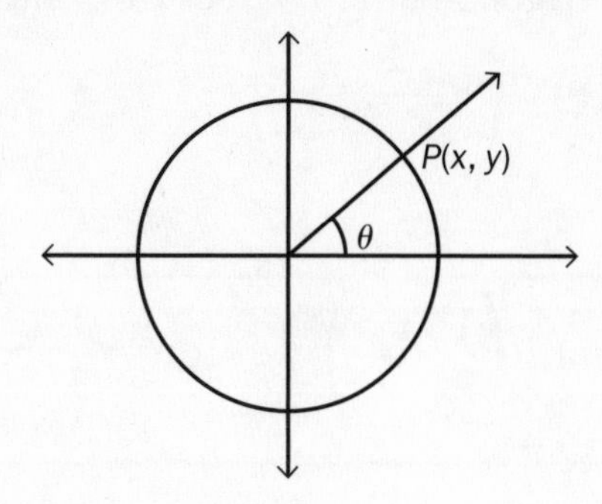

Figure 5.1

TRIGONOMETRIC RATIOS

We define the six trigonometric ratios in the unit circle ($r = 1$) as follows:

sine	$\sin\theta = y$, the value of the y-coordinate
cosine	$\cos\theta = x$, the value of the x-coordinate
tangent	$\tan\theta = \frac{y}{x}, x \neq 0$
cotangent	$\cot\theta = \frac{x}{y}, y \neq 0$
secant	$\sec\theta = \frac{1}{x}, x \neq 0$
cosecant	$\csc\theta = \frac{1}{y}, y \neq 0$

Note the reciprocal relationship between sine and cosecant; between cosine and secant; and between tangent and cotangent. If the point $P(x, y)$ is on the x-axis (so that $y = 0$), the cotangent and cosecant ratios are not defined. If the point $P(x, y)$ is on the y-axis (so that $x = 0$), the tangent and secant ratios are not defined.

Evaluating Trigonometric Ratios

Example

Find the six trigonometric values of the following points: (a) $P\left(\frac{1}{2}, \frac{\sqrt{3}}{2}\right)$; (b) $P\left(-\frac{\sqrt{2}}{2}, \frac{\sqrt{2}}{2}\right)$

Solution

(a) Given the point $P\left(\frac{1}{2}, \frac{\sqrt{3}}{2}\right)$, we see $x = \frac{1}{2}$ and $y = \frac{\sqrt{3}}{2}$. Therefore,

$$\sin\theta = y = \frac{\sqrt{3}}{2} \qquad \csc\theta = \frac{1}{y} = \frac{1}{\frac{\sqrt{3}}{2}} = \frac{2}{\sqrt{3}} = \frac{2\sqrt{3}}{3}$$

$$\cos\theta = x = \frac{1}{2} \qquad \sec\theta = \frac{1}{x} = \frac{1}{\frac{1}{2}} = 2$$

$$\tan\theta = \frac{y}{x} = \frac{\frac{\sqrt{3}}{2}}{\frac{1}{2}} = \sqrt{3} \qquad \cot\theta = \frac{x}{y} = \frac{\frac{1}{2}}{\frac{\sqrt{3}}{2}} = \frac{1}{\sqrt{3}} = \frac{\sqrt{3}}{3}$$

(b) Given the point $P\left(-\frac{\sqrt{2}}{2}, \frac{\sqrt{2}}{2}\right)$, we see $x = -\frac{\sqrt{2}}{2}$ and $y = \frac{\sqrt{2}}{2}$. Therefore,

$$\sin\theta = y = \frac{\sqrt{2}}{2} \qquad \csc\theta = \frac{1}{y} = \frac{1}{\frac{\sqrt{2}}{2}} = \frac{2}{\sqrt{2}} = \sqrt{2}$$

$$\cos\theta = x = -\frac{\sqrt{2}}{2} \qquad \sec\theta = \frac{1}{x} = \frac{1}{-\frac{\sqrt{2}}{2}} = -\frac{2}{\sqrt{2}} = -\sqrt{2}$$

$$\tan\theta = \frac{y}{x} = \frac{\frac{\sqrt{2}}{2}}{-\frac{\sqrt{2}}{2}} = -1 \qquad \cot\theta = \frac{x}{y} = \frac{-\frac{\sqrt{2}}{2}}{\frac{\sqrt{2}}{2}} = -1$$

The signs of the trigonometric functions are determined by the quadrants in which the points lie. The sine and cosecant functions are positive where y is positive, in the first and second quadrants. The cosine and secant functions are positive where x is positive, in the first and fourth quadrants. The tangent and cotangent functions are positive where x and y have the same sign—in the first quadrant, where x and y are both positive, and in the third quadrant, where x and y are both negative.

We can define the trigonometric ratios associated with various angles. A *radian* is defined as the measure of an angle with its vertex at the center of a circle and its arc equal in length to the radius of the circle. Since the circumference of a circle with radius 1 is $C = 2\pi r = 2\pi$, we can see that there are 2π radians in any circle, which is also 360°, so $360° = 2\pi$ radians. Thus, $1° = \frac{2\pi}{360} = \frac{\pi}{180}$ radians. Radians, the preferred angle measure for mathematics above the elementary level, are often expressed in terms of π. The designation "radians" is often omitted.

Each point P on the circle has a unique angle θ associated with it. For $\theta = 0 = 0°$, $P(x, y) = P(1, 0)$ is the corresponding point on the unit circle, and the six trigonometric values are:

$$\sin 0 = y = 0 \qquad \csc 0 = \frac{1}{y} = \frac{1}{0} = \textit{undefined}$$

$$\cos 0 = x = 1 \qquad \sec 0 = \frac{1}{x} = \frac{1}{1} = 1$$

$$\tan 0 = \frac{y}{x} = \frac{0}{1} = 0 \qquad \cot 0 = \frac{x}{y} = \frac{1}{0} = \textit{undefined}$$

For $\theta = \frac{\pi}{2} = 90°$, $P(x, y) = P(0, 1)$ is the corresponding point on the unit circle, and the six trigonometric values are:

$$\sin\frac{\pi}{2} = y = 1 \qquad \csc\frac{\pi}{2} = \frac{1}{y} = \frac{1}{1} = 1$$

$\cos\frac{\pi}{2} = x = 0$ $\sec\frac{\pi}{2} = \frac{1}{x} = \frac{1}{0} = \textit{undefined}$

$\tan\frac{\pi}{2} = \frac{y}{x} = \frac{1}{0} = \textit{undefined}$ $\cot\frac{\pi}{2} = \frac{x}{y} = \frac{0}{1} = 0$

For $\theta = \pi = 180°$, $P(x, y) = P(-1, 0)$ is the corresponding point on the unit circle, and the six trigonometric values are:

$\sin\pi = y = 0$ $\csc\pi = \frac{1}{y} = \frac{1}{0} = \textit{undefined}$

$\cos\pi = x = -1$ $\sec\pi = \frac{1}{x} = \frac{1}{-1} = -1$

$\tan\pi = \frac{y}{x} = \frac{0}{-1} = 0$ $\cot\pi = \frac{x}{y} = \frac{-1}{0} = \textit{undefined}$

For $\theta = \frac{3\pi}{2} = 270°$, $P(x, y) = P(0, -1)$ is the corresponding point on the unit circle, and the six trigonometric values are:

$\sin\frac{3\pi}{2} = y = -1$ $\csc\frac{3\pi}{2} = \frac{1}{y} = \frac{1}{-1} = -1$

$\cos\frac{3\pi}{2} = x = 0$ $\sec\frac{3\pi}{2} = \frac{1}{x} = \frac{1}{0} = \textit{undefined}$

$\tan\frac{3\pi}{2} = \frac{y}{x} = \frac{-1}{0} = \textit{undefined}$ $\cot\frac{3\pi}{2} = \frac{x}{y} = \frac{0}{-1} = 0$

Trigonometric values of additional angles can be found by using known ratios of sides in certain right triangles. For example, we know from geometry that a 45°-45°-90° triangle has side ratios of $1 : 1 : \sqrt{2}$, respectively. We can use this triangle information in our circle to determine the trigonometric ratios for $\frac{\pi}{4}(45°)$. Given the ratios of $1 : 1 : \sqrt{2}$ and the fact that the hypotenuse of the triangle is the radius of the unit circle ($r = 1$), we need to divide everything by $\sqrt{2}$ to have the new ratios of

$\frac{1}{\sqrt{2}} : \frac{1}{\sqrt{2}} : 1$, representing the ratios of the sides of our triangle in the unit circle. This is shown in Figure 5.2.

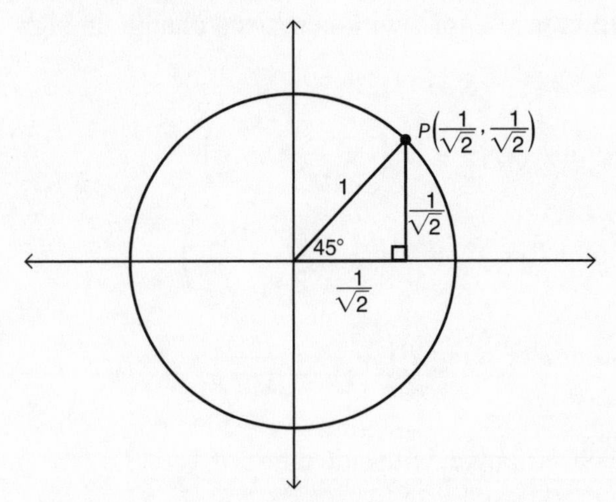

Figure 5.2

For $\theta = \frac{\pi}{4}$, $P(x, y) = P\left(\frac{1}{\sqrt{2}}, \frac{1}{\sqrt{2}}\right)$ is the corresponding point on the unit circle, and the six trigonometric values are:

$$\sin\frac{\pi}{4} = y = \frac{1}{\sqrt{2}} \qquad \csc\frac{\pi}{4} = \frac{1}{y} = \frac{1}{\frac{1}{\sqrt{2}}} = \sqrt{2}$$

$$\cos\frac{\pi}{4} = x = \frac{1}{\sqrt{2}} \qquad \sec\frac{\pi}{4} = \frac{1}{x} = \frac{1}{\frac{1}{\sqrt{2}}} = \sqrt{2}$$

$$\tan\frac{\pi}{4} = \frac{y}{x} = \frac{\frac{1}{\sqrt{2}}}{\frac{1}{\sqrt{2}}} = 1 \qquad \cot\frac{\pi}{4} = \frac{x}{y} = \frac{\frac{1}{\sqrt{2}}}{\frac{1}{\sqrt{2}}} = 1$$

We can also determine the values of angles that have $\frac{\pi}{4}$ as a reference angle. A *reference angle* is defined as the positive acute angle formed between the terminal side of an angle and the x-axis. So, for example, $\frac{\pi}{4}$ would be a reference angle for $\pi - \frac{\pi}{4}$ in the second quadrant, $\pi + \frac{\pi}{4}$ in the third quadrant, and $2\pi - \frac{\pi}{4}$ in the fourth quadrant. The values of the trigonometric functions of these angles are the same as the trigonometric values of the reference angle, with perhaps a change in sign, depending on the quadrant.

$\frac{3\pi}{4}$ has point $P(x, y) = P\left(-\frac{1}{\sqrt{2}}, \frac{1}{\sqrt{2}}\right)$

$\frac{5\pi}{4}$ has point $P(x, y) = P\left(-\frac{1}{\sqrt{2}}, -\frac{1}{\sqrt{2}}\right)$

$\frac{7\pi}{4}$ has point $P(x, y) = P\left(\frac{1}{\sqrt{2}}, -\frac{1}{\sqrt{2}}\right)$

A 30°-60°-90° triangle, with side ratios of $1 : \sqrt{3} : 2$, respectively, can be used to determine the trigonometric ratios for $\frac{\pi}{6}$(30°) if the 30° angle is the central angle and for $\frac{\pi}{3}$(60°) if the central angle measures 60°. The trigonometric ratios of all angles with $\frac{\pi}{6}$ and $\frac{\pi}{3}$ as reference angles can be found using these ratios, remembering to use the signs associated with the coordinates of the points in the different quadrants. Given the ratio of $1 : \sqrt{3} : 2$ and the fact that the hypotenuse of the triangle is the radius of the unit circle ($r = 1$), we need to divide the ratios by 2 to have the new ratios of $\frac{1}{2} : \frac{\sqrt{3}}{2} : 1$, representing the ratios of the sides of our triangle in the unit circle. See Figures 5.3 and 5.4 for $\frac{\pi}{6}$ and $\frac{\pi}{3}$, respectively.

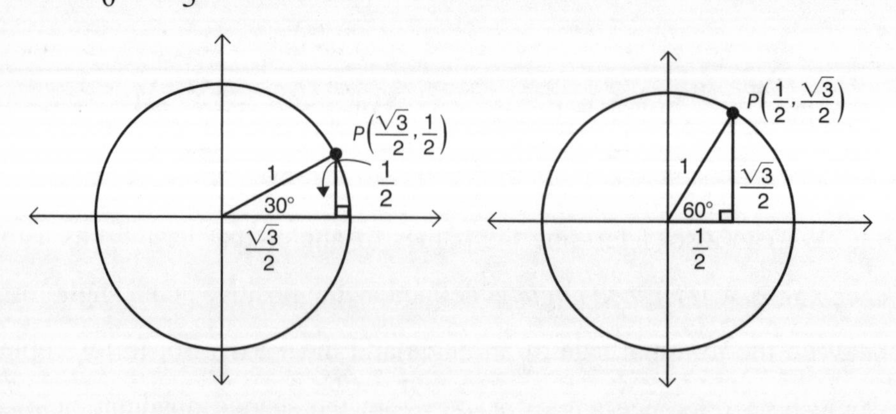

Figure 5.3 **Figure 5.4**

For $\theta = \frac{\pi}{6}$, $P(x, y) = P\left(\frac{\sqrt{3}}{2}, \frac{1}{2}\right)$ is the corresponding point on the unit circle, and the six trigonometric values are:

$$\sin\frac{\pi}{6} = y = \frac{1}{2} \qquad \csc\frac{\pi}{6} = \frac{1}{y} = \frac{1}{\frac{1}{2}} = 2$$

$$\cos\frac{\pi}{6} = x = \frac{\sqrt{3}}{2} \qquad \sec\frac{\pi}{6} = \frac{1}{x} = \frac{1}{\frac{\sqrt{3}}{2}} = \frac{2}{\sqrt{3}}$$

$$\tan\frac{\pi}{6} = \frac{y}{x} = \frac{\frac{1}{2}}{\frac{\sqrt{3}}{2}} = \frac{1}{\sqrt{3}} \qquad \cot\frac{\pi}{6} = \frac{x}{y} = \frac{\frac{\sqrt{3}}{2}}{\frac{1}{2}} = \sqrt{3}$$

We can also determine the values of the angles that have $\frac{\pi}{6}$ as a reference angle.

$\frac{5\pi}{6}$ with the point $P(x, y) = P\left(-\frac{\sqrt{3}}{2}, \frac{1}{2}\right)$

$\frac{7\pi}{6}$ with the point $P(x, y) = P\left(-\frac{\sqrt{3}}{2}, -\frac{1}{2}\right)$

$\frac{11\pi}{6}$ with the point $P(x, y) = P\left(\frac{\sqrt{3}}{2}, -\frac{1}{2}\right)$

For $\theta = \frac{\pi}{3}$, $P(x, y) = P\left(\frac{1}{2}, \frac{\sqrt{3}}{2}\right)$ is the corresponding point on the unit circle, and the six trigonometric values are:

$$\sin\frac{\pi}{3} = y = \frac{\sqrt{3}}{2} \qquad \csc\frac{\pi}{3} = \frac{1}{y} = \frac{1}{\frac{\sqrt{3}}{2}} = \frac{2}{\sqrt{3}}$$

$$\cos\frac{\pi}{3} = x = \frac{1}{2} \qquad \sec\frac{\pi}{3} = \frac{1}{x} = \frac{1}{\frac{1}{2}} = 2$$

$$\tan\frac{\pi}{3}=\frac{y}{x}=\frac{\frac{\sqrt{3}}{2}}{\frac{1}{2}}=\sqrt{3} \qquad \cot\frac{\pi}{3}=\frac{x}{y}=\frac{\frac{1}{2}}{\frac{\sqrt{3}}{2}}=\frac{1}{\sqrt{3}}$$

We can also determine the values of the angles that have $\frac{\pi}{3}$ as a reference angle.

$\frac{2\pi}{3}$ with the point $P(x, y) = P\left(-\frac{1}{2}, \frac{\sqrt{3}}{2}\right)$

$\frac{4\pi}{3}$ with the point $P(x, y) = P\left(-\frac{1}{2}, -\frac{\sqrt{3}}{2}\right)$

$\frac{5\pi}{3}$ with the point $P(x, y) = P\left(\frac{1}{2}, -\frac{\sqrt{3}}{2}\right)$

Note: Many times the denominators of the fractions with radicals are rationalized so that we would use $\frac{\sqrt{2}}{2}$, $\frac{2\sqrt{3}}{3}$ and $\frac{\sqrt{3}}{3}$ instead of $\frac{1}{\sqrt{2}}$, $\frac{2}{\sqrt{3}}$, and $\frac{1}{\sqrt{3}}$, respectively.

Circle with a Radius Other Than 1

We have been using a unit circle ($r = 1$), but we can generalize to a circle with any positive radius r. Given θ in standard position and $P(x, y)$ a point on the circle $x^2 + y^2 = r^2$, we can define

$$\sin\theta = \frac{y}{r} \qquad \csc\theta = \frac{r}{y},\ y \neq 0$$

$$\cos\theta = \frac{x}{r} \qquad \sec\theta = \frac{r}{x},\ x \neq 0$$

$$\tan\theta = \frac{y}{x},\ x \neq 0 \qquad \cot\theta = \frac{x}{y},\ y \neq 0$$

Arc Length

Another measure we have in a circle is that of arc length. The length of an arc of a circle is given by the formula $s = r\theta$, where s is the arc length, r is the radius, and θ is the central angle, measured in radians, that forms the arc.

Right Triangle Trigonometry

We can also define the trigonometric ratios outside a circle by using only a right triangle. Given a right triangle with acute angle θ, we define the ratios

$$\sin\theta = \frac{opp}{hyp} \qquad \csc\theta = \frac{hyp}{opp} = \frac{1}{\sin\theta}$$

$$\cos\theta = \frac{adj}{hyp} \qquad \sec\theta = \frac{hyp}{adj} = \frac{1}{\cos\theta}$$

$$\tan\theta = \frac{opp}{adj} = \frac{\sin\theta}{\cos\theta} \qquad \cot\theta = \frac{adj}{opp} = \frac{1}{\tan\theta}$$

where "adj" is the side adjacent to the angle θ, "opp" is the side opposite θ, and "hyp" is the hypotenuse of the triangle.

Example

Find the remaining trigonometric values if $\tan\theta = \frac{1}{2}$ and $\sin\theta < 0$.

Solution

Since $\tan\theta = \frac{y}{x}$ and we are given $\tan\theta = \frac{1}{2}$, then we know $\frac{y}{x} = \frac{1}{2}$. Since $x^2 + y^2 = r^2$, then $2^2 + 1^2 = r^2$ implies that $r = \sqrt{5}$. Also, since tangent is given as positive and sine as negative, this tells us that the angle is in the third quadrant so x and y are both negative. Hence, $x = -2$ and $y = -1$. Using these values and the trigonometric ratios, we have

$$\sin\theta = \frac{y}{r} = \frac{-1}{\sqrt{5}} \qquad \csc\theta = \frac{r}{y} = \frac{\sqrt{5}}{-1} = -\sqrt{5}$$

$$\cos\theta = \frac{x}{r} = \frac{-2}{\sqrt{5}} \qquad \sec\theta = \frac{r}{x} = \frac{\sqrt{5}}{-2}$$

$$\tan\theta = \frac{y}{x} = \frac{-1}{-2} = \frac{1}{2} \qquad \cot\theta = \frac{x}{y} = \frac{-2}{-1} = 2$$

TRIGONOMETRIC IDENTITIES

Consider points $P(x, y)$ on the circle $x^2 + y^2 = r^2$, with θ as the reference angle. Given the standard trigonometric ratios:

$$\sin\theta = \frac{y}{r} \qquad \csc\theta = \frac{r}{y},\ y \neq 0$$

$$\cos\theta = \frac{x}{r} \qquad \sec\theta = \frac{r}{x},\ x \neq 0$$

$$\tan\theta = \frac{y}{x},\ x \neq 0 \qquad \cot\theta = \frac{x}{y},\ y \neq 0$$

we can use algebra to prove several trigonometric identities. The solutions to the following examples prove some of these.

Example

Prove (a) $\sin^2\theta + \cos^2\theta = 1$; (b) $\tan^2\theta + 1 = \sec^2\theta$; and (c) $\cot^2\theta + 1 = \csc^2\theta$.

Solution

(a) $\sin^2\theta + \cos^2\theta = 1$

$$\begin{aligned}\sin^2\theta + \cos^2\theta &= \left(\frac{y}{r}\right)^2 + \left(\frac{x}{r}\right)^2 \\ &= \frac{y^2 + x^2}{r^2} \\ &= \frac{r^2}{r^2} \\ &= 1\end{aligned}$$

(b) $\tan^2\theta + 1 = \sec^2\theta$

$$\begin{aligned}\tan^2\theta + 1 &= \left(\frac{y}{x}\right)^2 + 1 \\ &= \frac{y^2}{x^2} + \frac{x^2}{x^2} \\ &= \frac{y^2 + x^2}{x^2}\end{aligned}$$

$$= \frac{r^2}{x^2}$$

$$= \sec^2\theta$$

(c) $\cot^2\theta + 1 = \csc^2\theta$

$$\cot^2\theta + 1 = \left(\frac{x}{y}\right)^2 + 1$$

$$= \frac{x^2}{y^2} + \frac{y^2}{y^2}$$

$$= \frac{x^2 + y^2}{y^2}$$

$$= \frac{r^2}{y^2}$$

$$= \csc^2\theta$$

There are many other trigonometric identities that we can prove. Remember that an identity is a mathematical statement that is true for all values of the variable.

Example

Prove $\frac{\sin^3 x}{\cos x} + \sin x \cos x = \tan x$.

Solution

We show that the left side of the identity is equal to $\tan x$. First, we factor the numerator.

$$\frac{\sin x(\sin^2 x)}{\cos x} + \sin x \cos x$$

Using the identity $\sin^2 x + \cos^2 x = 1$ from the last example, we have

$$\frac{\sin x(1 - \cos^2 x)}{\cos x} + \sin x \cos x =$$

$$\sin x\left[\frac{1 - \cos^2 x}{\cos x} + \cos x\right] =$$

$$\sin x\left[\frac{1}{\cos x}\right] = \tan x$$

Example

Prove $\dfrac{\sin x - 1}{\cos x} - \dfrac{\cos x}{\sin x - 1} = 2\tan x.$

Solution

We show that the left side of the identity is equal to 2 tan x.

$$\frac{(\sin x - 1)(\sin x - 1)}{(\cos x)(\sin x - 1)} - \frac{\cos x(\cos x)}{(\sin x - 1)(\cos x)} =$$

$$\frac{(\sin^2 x - 2\sin x + 1)}{(\cos x)(\sin x - 1)} - \frac{\cos x(\cos x)}{(\sin x - 1)(\cos x)} =$$

$$\frac{\sin^2 x - 2\sin x + 1 - \cos^2 x}{\cos x(\sin x - 1)} =$$

$$\frac{\sin^2 x - 2\sin x + (\sin^2 x + \cos^2 x) - \cos^2 x}{\cos x(\sin x - 1)} =$$

$$\frac{2\sin^2 x - 2\sin x}{\cos x(\sin x - 1)} =$$

$$\frac{2\sin x(\sin x - 1)}{\cos x(\sin x - 1)} =$$

$$\frac{2\sin x}{\cos x} = 2\tan x$$

SKETCHING THE TRIGONOMETRIC FUNCTIONS

If $f(\theta) = \sin\theta$, then $f(\theta) = \sin(\theta + 2\pi k) = \sin\theta$, where k is any integer. If $f(\theta) = \cos\theta$, then $f(\theta) = \cos(\theta + 2\pi k) = \cos\theta$, where k is any integer. This is true for the other trigonometric functions also. But we want to find the smallest values of p so that $f(\theta + p) = f(\theta)$. Functions of this type that repeat at regular intervals are called *periodic functions*, and the smallest value p for which this is true is called the *period* of the function. The period of sine, cosine, cosecant, and secant functions is 2π. The period of tangent and cotangent is π.

We recall that if $f(-\theta) = f(\theta)$, the function f is even and if $f(-\theta) = -f(\theta)$, the function f is odd for all θ in the domain.

$y = \sin\theta$

Domain: $(-\infty, \infty)$

Range: $[-1, 1]$

Period: 2π

Symmetry: Odd, since $\sin(-\theta) = -\sin\theta$

Sketch:

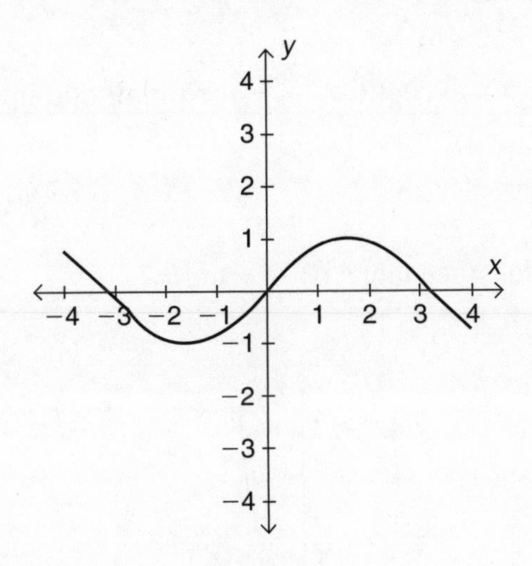

Figure 5.5

$y = \cos\theta$

Domain: $(-\infty, \infty)$

Range: $[-1, 1]$

Period: 2π

Symmetry: Even, since $\cos(-\theta) = \cos\theta$

Sketch:

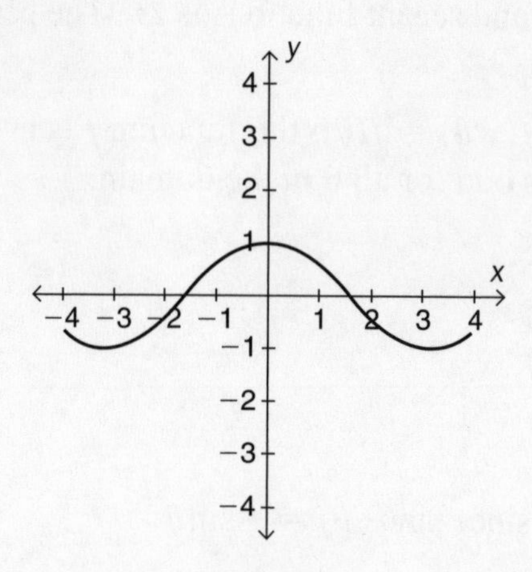

Figure 5.6

$\boldsymbol{y = \tan\theta}$

Domain: $(-\infty, \infty)$, except $\theta \neq \frac{\pi}{2} + n\pi$ for all odd integers n

Range: $(-\infty, \infty)$

Period: π

Symmetry: Odd, since $\tan(-\theta) = -\tan\theta$

Sketch:

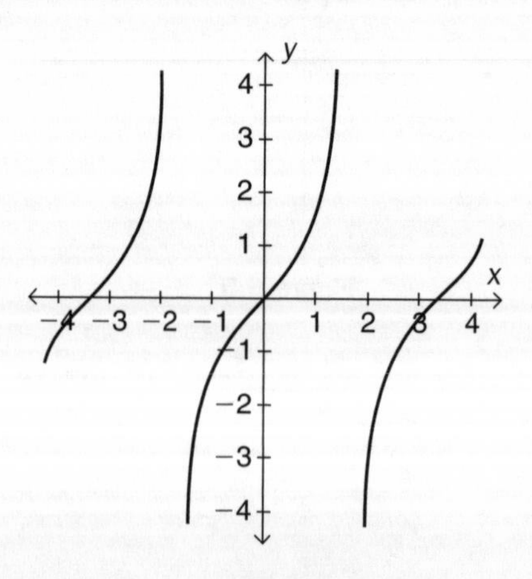

Figure 5.7

$\boldsymbol{y = \cot\theta}$

Domain: $(-\infty, \infty)$, except $\theta \neq n\pi$ for all integers n

Range: $(-\infty, \infty)$

Period: π

Symmetry: Odd, since $\cot(-\theta) = -\cot\theta$

Sketch:

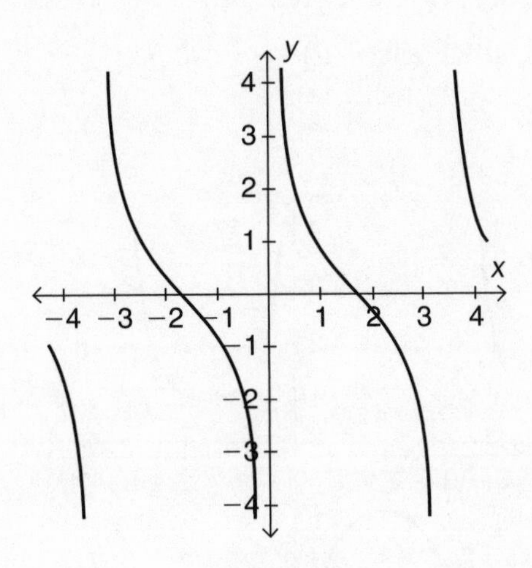

Figure 5.8

y* = sec *θ

Domain: $(-\infty, \infty)$, except $\theta \neq \dfrac{\pi}{2} + n\pi$ for all odd integers n

Range: $|y| \geq 1$

Period: 2π

Symmetry: Even, since $\sec(-\theta) = \sec\theta$

Sketch:

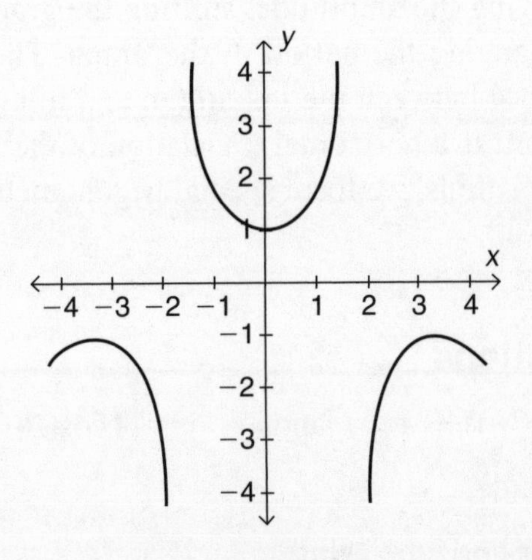

Figure 5.9

$y = \csc\theta$

Domain: $(-\infty, \infty)$, except $\theta \neq n\pi$ for all integers n

Range: $|y| \geq 1$

Period: 2π

Symmetry: Odd, since $\csc(-\theta) = -\csc\theta$

Sketch:

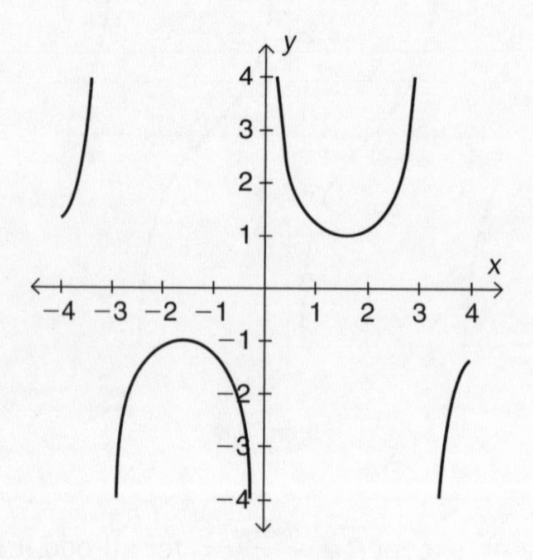

Figure 5.10

TRANSFORMATIONS

We can find variations of the trigonometric functions by using transformations such as varying the amplitude, shifting the graph horizontally and vertically, and/or changing the period of the graph. The amplitude is half of the vertical distance between the maximum and minimum values of the function. A phase shift is a horizontal translation of the function. When we make these transformations, we form special trigonometric functions called *sinusoidal functions*.

Sinusoidal Functions

$y = A\sin(B(\theta - h)) + k$ and $y = A\cos(B(\theta - h)) + k$

$|A|$ is the amplitude.

h is the phase or horizontal shift.

$y = k$ is the horizontal line through the middle of the graph and determines a vertical shift.

$|B|$ is the number of cycles completed in 2π radians.

$\frac{2\pi}{|B|}$ is the period of the graph.

Example

What is the minimum value of $y = \frac{1}{2}\sin 3x - \frac{3}{2}$?

Solution

We see that the amplitude is $\frac{1}{2}$. We also notice that $k = -\frac{3}{2}$ in the equation. Therefore, the line $y = -\frac{3}{2}$ is the line through the middle of the graph. Since the amplitude is $\frac{1}{2}$, we only need to go down $\frac{1}{2}$ units from the midline to get the minimum value. Hence, the minimum value is $-\frac{3}{2} - \frac{1}{2} = -2$.

Example

Analyze and sketch $y = \frac{5}{2}\sin 2\left(x + \frac{\pi}{2}\right) + 1$.

Solution

Amplitude: $\frac{5}{2}$

Period: $\frac{2\pi}{2} = \pi$

Vertical shift: Up one unit

Phase or horizontal shift: $\frac{\pi}{2}$ units to the left, since we can write $\left(x + \frac{\pi}{2}\right)$ as $\left(x - \left(-\frac{\pi}{2}\right)\right)$

From the parent function $y = \sin x$, we change the amplitude to $\frac{5}{2}$, shift the graph $\frac{\pi}{2}$ units to the left, squeeze the graph so that the new period is π, and raise the graph one unit. The resulting sketch is shown in Figure 5.11.

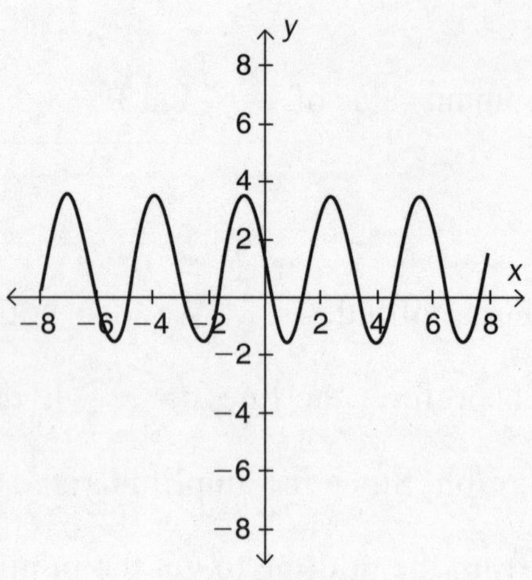

Figure 5.11

Example

What is the phase (or horizontal) shift of $y = \frac{2}{3}\sin\left(x + \frac{\pi}{3}\right) - 1$?

Solution

We can write the equation as $y = \frac{2}{3}\sin\left(x - \left(-\frac{\pi}{3}\right)\right) - 1$ and determine that the phase shift is $\frac{\pi}{3}$ units to the left.

Example

What is the period of the function $y = 3\sin\left(\frac{2\pi}{3}x - 1\right) + \frac{3}{4}$?

Solution

From the general form of a sinusoidal function, $y = A\sin(B(\theta - h)) + k$, we know that the period of the function is $\frac{2\pi}{|B|}$. We can write this function in the general form for a sinusoidal function, $y = 3\sin\left(\frac{2\pi}{3}\left(x - \frac{3}{2\pi}\right)\right) + \frac{3}{4}$ where $B = \frac{2\pi}{3}$. Hence, the period is $\frac{2\pi}{|B|} = \frac{2\pi}{\left(\frac{2\pi}{3}\right)} = 3$.

INVERSE TRIGONOMETRIC FUNCTIONS

Notice that the trigonometric functions are not 1-1 over their domains and therefore do not have inverses that are functions. But we can fix this problem by restricting the domains of the six trigonometric functions so that their inverses are functions. We need to restrict the domain so that the entire range is included but is 1-1 over this restricted domain.

Notice that the domains of each function can be restricted in the following manner:

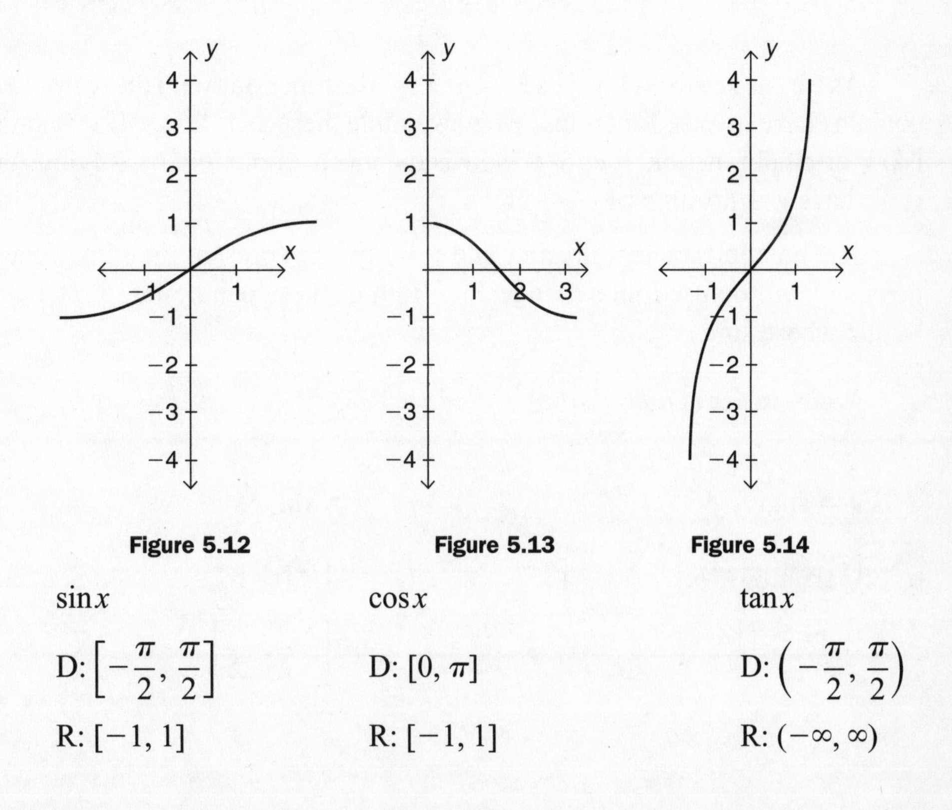

Figure 5.12 **Figure 5.13** **Figure 5.14**

$\sin x$	$\cos x$	$\tan x$
D: $\left[-\frac{\pi}{2}, \frac{\pi}{2}\right]$	D: $[0, \pi]$	D: $\left(-\frac{\pi}{2}, \frac{\pi}{2}\right)$
R: $[-1, 1]$	R: $[-1, 1]$	R: $(-\infty, \infty)$

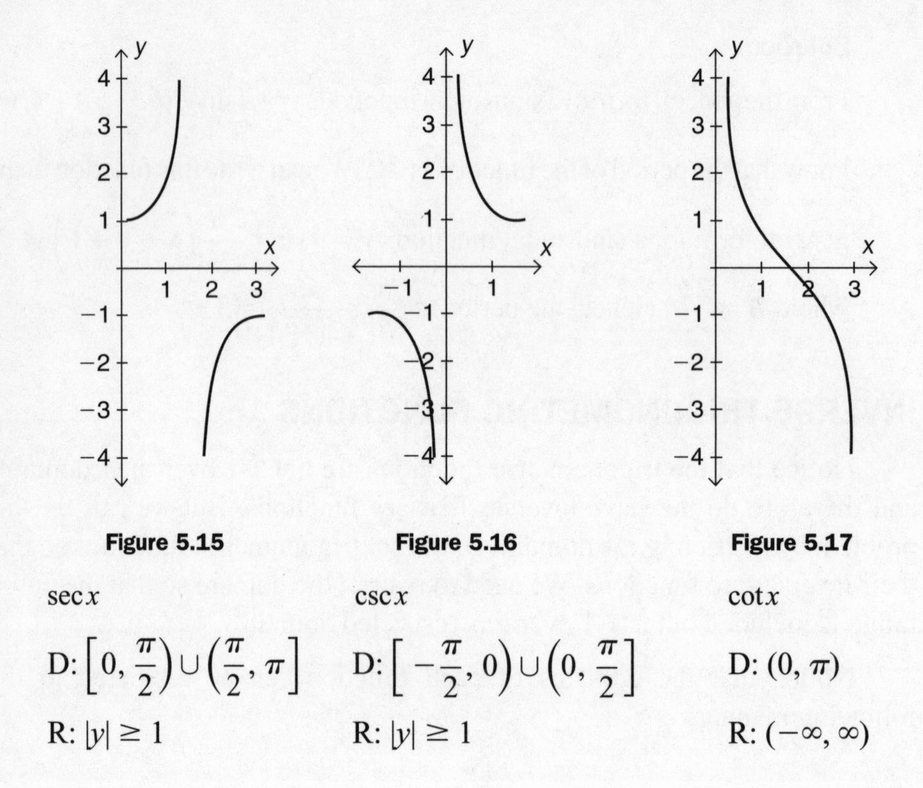

Figure 5.15 | Figure 5.16 | Figure 5.17

$\sec x$

D: $\left[0, \frac{\pi}{2}\right) \cup \left(\frac{\pi}{2}, \pi\right]$

R: $|y| \geq 1$

$\csc x$

D: $\left[-\frac{\pi}{2}, 0\right) \cup \left(0, \frac{\pi}{2}\right]$

R: $|y| \geq 1$

$\cot x$

D: $(0, \pi)$

R: $(-\infty, \infty)$

As we can see from Figures 5.12 to 5.17, the functions with the restricted domains have inverse functions. We now define these as follows: If $y = \sin x$ is our original function, then $x = \sin y$ is the inverse function, or, solving for y, we have $y = \arcsin x$ or $y = \sin^{-1} x$.

The notations of $y = \arcsin x$ and $y = \sin^{-1} x$ represent the same functions and will be used interchangeably. Both of these can be interpreted as "angle whose sine is x."

Note: $\sin^{-1} x$ is *not* $\frac{1}{\sin x}$!

$y = \sin x$	$y = \sin^{-1} x$
D (restricted): $\left[-\frac{\pi}{2}, \frac{\pi}{2}\right]$	D: $[-1, 1]$
R: $[-1, 1]$	R: $\left[-\frac{\pi}{2}, \frac{\pi}{2}\right]$

$\boldsymbol{y = \cos x}$

D (restricted): $[0, \pi]$

R: $[-1, 1]$

$\boldsymbol{y = \cos^{-1} x}$

D: $[-1, 1]$

R: $[0, \pi]$

$\boldsymbol{y = \tan x}$

D (restricted): $\left(-\frac{\pi}{2}, \frac{\pi}{2}\right)$

R: $(-\infty, \infty)$

$\boldsymbol{y = \tan^{-1} x}$

D: $(-\infty, \infty)$

R: $\left(-\frac{\pi}{2}, \frac{\pi}{2}\right)$

$\boldsymbol{y = \cot x}$

D (restricted): $(0, \pi)$

R: $(-\infty, \infty)$

$\boldsymbol{y = \cot^{-1} x}$

D: $(-\infty, \infty)$

R: $(0, \pi)$

$\boldsymbol{y = \sec x}$

D (restricted): $\left[0, \frac{\pi}{2}\right) \cup \left(\frac{\pi}{2}, \pi\right]$

R: $|y| \geq 1$

$\boldsymbol{y = \sec^{-1} x}$

D: $|x| \geq 1$

R: $\left[0, \frac{\pi}{2}\right) \cup \left(\frac{\pi}{2}, \pi\right]$

$\boldsymbol{y = \csc x}$

D (restricted): $\left[-\frac{\pi}{2}, 0\right) \cup \left(0, \frac{\pi}{2}\right]$

R: $|y| \geq 1$

$\boldsymbol{y = \csc^{-1} x}$

D: $|x| \geq 1$

R: $\left[-\frac{\pi}{2}, 0\right) \cup \left(0, \frac{\pi}{2}\right]$

Example

Find the exact values of (a) $\sin\left(\tan^{-1}\left(\frac{1}{2}\right)\right)$ and

(b) $\cos\left(\sin^{-1}\left(-\frac{1}{3}\right)\right)$.

Solution

(a) Let $\theta = \tan^{-1}\frac{1}{2}$. Because of the restrictions on θ for the inverse tangent function, we know $-\frac{\pi}{2} < \theta < \frac{\pi}{2}$. Since $\tan\theta$ is positive in this problem $\left(\frac{1}{2}\right)$, then θ must be in the first quadrant. We know $\tan\theta = \frac{y}{x}$ and $\tan\theta = \frac{1}{2}$ (given). Therefore, $\frac{y}{x} = \frac{1}{2}$. Using the fact that $x^2 + y^2 = r^2$, we find that $r = \sqrt{5}$. Hence, $\sin\theta = \frac{y}{r} = \frac{1}{\sqrt{5}}$.

(b) Let $\theta = \sin^{-1}\left(-\frac{1}{3}\right)$. Because of the restrictions on θ for the inverse sine function, we know $-\frac{\pi}{2} \leq \theta \leq \frac{\pi}{2}$. Since $\sin\theta$ is given in this problem as negative, then θ must be in the fourth quadrant. Since $\sin\theta = \frac{y}{r}$ and $\sin\theta = \frac{-1}{3}$ (given), then we know $\frac{y}{r} = \frac{-1}{3}$. Using the fact that $x^2 + y^2 = r^2$, we can find that $x = 2\sqrt{2}$. Therefore, $\cos\theta = \frac{x}{r} = \frac{2\sqrt{2}}{3}$. It is positive, since our angle was in the fourth quadrant and cosine is positive in this quadrant.

Example

Write the following as an expression in x: $\cos(\tan^{-1}x)$.

Solution

Let $\theta = \tan^{-1} x$, or rewritten as $\tan\theta = x$. Since $\tan\theta = \frac{opp}{adj} = x = \frac{x}{1}$, we can make assignments to the sides of a triangle.

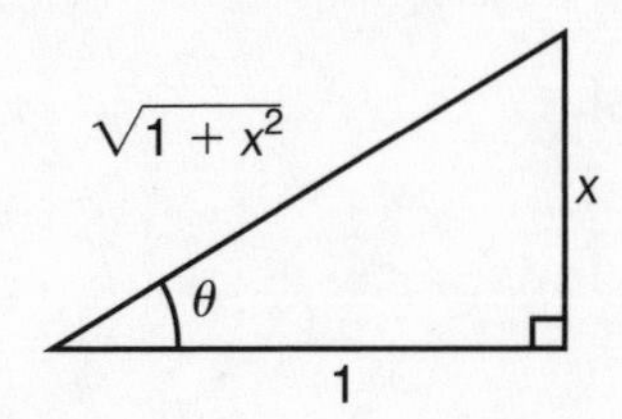

We use the Pythagorean Theorem to find that the hypotenuse is $\sqrt{1+x^2}$. Therefore, we have $\cos(\tan^{-1}x) = \cos\theta = \frac{adj}{hyp} = \frac{1}{\sqrt{1+x^2}}$.

Example

What is the domain of $y = 3\sin^{-1}(2x)$?

Solution

The domain for the arcsine function is $[-1, 1]$. Therefore,

$-1 < 2x < 1$ or, solving for x,

$-\frac{1}{2} \leq x \leq \frac{1}{2}$

Example

Solve (a) $3\sin^{-1} x = \frac{\pi}{2}$ and (b) $3\cos^{-1}(2x) = 2\pi$ for x.

Solution

(a) $$3\sin^{-1} x = \frac{\pi}{2}$$

$$\sin^{-1} x = \frac{\pi}{6}$$

$$x = \sin\frac{\pi}{6}$$

$$x = \frac{1}{2}$$

(b) $3\cos^{-1}(2x) = 2\pi$

$$\cos^{-1}(2x) = \frac{2\pi}{3}$$

$$2x = \cos\left(\frac{2\pi}{3}\right)$$

$$2x = -\frac{1}{2}$$

$$x = -\frac{1}{4}$$

TRIGONOMETRIC EQUATIONS

Example

Solve: $2(\cos x + 1) = 2$ for x.

Solution

$2(\cos x + 1) = 2$

Dividing by 2 and subtracting 1, we have

$\cos x + 1 = 1$

$\cos x = 0 \quad \Rightarrow x = \frac{\pi}{2} + k\pi$, where k is any integer

Example

Solve $\tan^2 3x = 3$ for x, $0 < x < \frac{\pi}{2}$.

Solution

Since $0 < x < \frac{\pi}{2}$, we can multiply by 3 to get $0 < 3x < \frac{3\pi}{2}$, since the angle is expressed as $3x$.

Since $\quad \tan^2 3x = 3$

then $\quad \tan 3x = \pm\sqrt{3}$

The angles where the tangent is equal to $\pm\sqrt{3}$ in the first three quadrants are $\frac{\pi}{3}, \frac{2\pi}{3}, \frac{4\pi}{3}$.

Therefore, $3x = \frac{\pi}{3}$, $3x = \frac{2\pi}{3}$, and $3x = \frac{4\pi}{3}$.

Or, solving for x: $x = \frac{\pi}{9}$, $x = \frac{2\pi}{9}$, and $x = \frac{4\pi}{9}$.

Example

Solve $2\cos^2 2x - 4\cos 2x = 1$ for x, $0 \le x < \frac{\pi}{2}$.

Solution

Since $0 \le x < \frac{\pi}{2}$, then $0 \le 2x < \pi$. We recognize that the original equation is a quadratic equation in $\cos(2x)$. Since it will not factor, we use the quadratic formula, with $a = 2$, $b = -4$, and $c = -1$.

$$\cos(2x) = \frac{4 \pm \sqrt{(-4)^2 - 4(2)(-1)}}{2(2)} = \frac{4 \pm \sqrt{24}}{4} = \frac{4 \pm 2\sqrt{6}}{4}$$

$$= \frac{2 \pm \sqrt{6}}{2} \approx -.2247, +2.225$$

We discard 2.225 since it is out of the range for cosine.

$\cos(2x) \approx -.2247$

or $\quad 2x = \cos^{-1}(-.2247) \approx 1.797$

$x \approx .8987$ radians

Notice that $2x$ is in the second quadrant, which is why our solution to $\cos(2x)$ is negative, but x is in the first quadrant, since we were given $0 \le x < \frac{\pi}{2}$.

OTHER TRIANGLES

The trigonometric ratios we have discussed so far apply to right triangles. We can generalize to all triangles, not just right triangles. Similar to the Pythagorean Theorem for right triangles, the Law of Cosines relates the sides of any triangle.

Law of Cosines: For a triangle with sides a, b, and c, with angle C opposite side c, we have

$$c^2 = a^2 + b^2 - 2ab \cos C$$

or equivalent forms

$$a^2 = b^2 + c^2 - 2bc \cos A$$

$$b^2 = a^2 + c^2 - 2ac \cos B$$

We can also use the following *Law of Sines* in solving for angles or lengths of sides in a triangle.

Law of Sines: For a triangle with sides a, b, and c, and opposite angles A, B, C, respectively,

$$\frac{\sin A}{a} = \frac{\sin B}{b} = \frac{\sin C}{c}$$

Example

Solve for angle C.

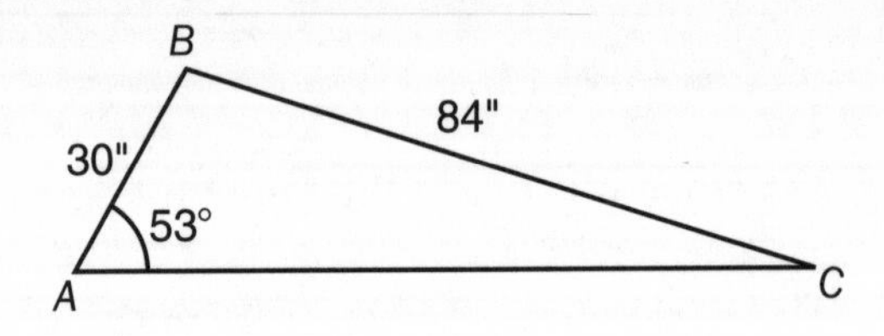

Solution

We can solve for angle C by using the Law of Sines.

$$\frac{\sin A}{a} = \frac{\sin C}{c}$$

$$\frac{\sin 53^\circ}{84} = \frac{\sin C}{30}$$

$$\sin C = \frac{30 \sin 53^\circ}{84} \approx .2852$$

$$C \approx \sin^{-1}(.2852) \approx 16.57^\circ$$

Example

A 45-foot ladder needs to be placed at what angle with the ground so that the top of the ladder will reach a window that is 22 feet above the ground?

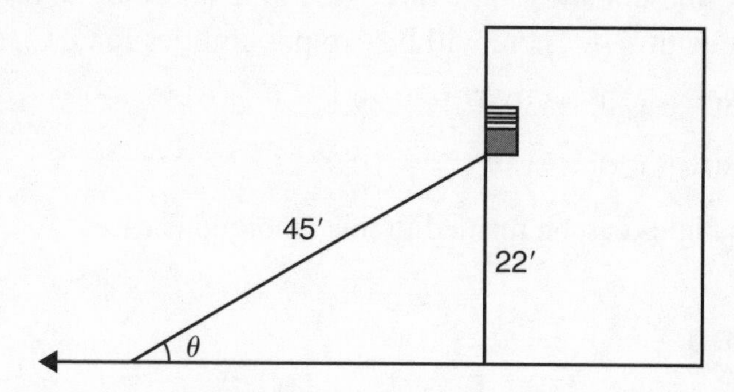

Solution

Since $\sin\theta = \frac{opp}{hyp}$, we have $\sin\theta = \frac{22}{45}$, or $\theta = \sin^{-1}\left(\frac{22}{45}\right)$.

$$\theta = \sin^{-1}\left(\frac{22}{45}\right) \approx .51 \text{ radians}$$

We now convert the radians to degrees.

$$(.51) \times \left(\frac{180}{\pi}\right) \approx 29.22°$$

If we are given two sides of a triangle and the angle opposite one of them, then these sides and angle may form one, two, or no triangles. This is considered an *ambiguous* case.

Example

Given $a = 4$, $b = 6$, and $\angle A = 20°$, find $\angle B$.

Solution

We use the Law of Sines.

$$\frac{\sin A}{a} = \frac{\sin B}{b}$$

$$\frac{\sin 20°}{4} = \frac{\sin B}{6}$$

$$\sin B = \frac{6 \sin 20^\circ}{4} \approx .513$$

$$B \approx 30.9^\circ \text{ or } B \approx 149.1^\circ$$

Both of these angles have this sine value and they both will form a triangle with $\angle A$. There will be two possibilities for $\angle C$.

$C \approx 180^\circ - (20^\circ + 30.9^\circ) \approx 129.1^\circ$

$C \approx 180^\circ - (20^\circ + 149.1^\circ) \approx 10.9^\circ$

Two triangles can be formed in this ambiguous case.

Example

Given $a = 2$, $b = 1$, and $\angle B = 45^\circ$. Find $\angle A$.

Solution

We use the Law of Sines.

$$\frac{\sin A}{a} = \frac{\sin B}{b}$$

$$\frac{\sin A}{2} = \frac{\sin 45^\circ}{1}$$

$$\sin A = \frac{2 \sin 45^\circ}{1} \approx 1.414$$

This value for sine is outside the range of $[-1, 1]$. These values for the sides and angle will not form a triangle in this ambiguous case.

Example

Given $a = 5$, $b = 8$, and $c = 9$. Find $\angle A$, $\angle B$, and $\angle C$.

Solution

Given three sides, use the Law of Cosines to find the angles.

$a^2 = b^2 + c^2 - 2bc \cos A$

$5^2 = 8^2 + 9^2 - 2(8)(9) \cos A$

$$25 = 64 + 81 - 144 \cos A$$

$$-120 = -144 \cos A$$

$$\cos A = \frac{120}{144} \approx .833$$

$$A \approx 33.6°$$

$$b^2 = a^2 + c^2 - 2ac \cos B$$

$$8^2 = 5^2 + 9^2 - 2(5)(9) \cos B$$

$$64 = 25 + 81 - 90 \cos B$$

$$-42 = -90 \cos B$$

$$\cos B = \frac{42}{90} \approx .467$$

$$B \approx 62.2°$$

Since $\angle A + \angle B + \angle C = 180°$, we have

$$33.6° + 62.2° + \angle C \approx 180°$$

$$\angle C \approx 84.2°$$

SUMMARY OF TRIGONOMETRIC IDENTITIES

Basic Identities

$$\tan\theta = \frac{\sin\theta}{\cos\theta} \qquad \cot\theta = \frac{\cos\theta}{\sin\theta} \qquad \csc\theta = \frac{1}{\sin\theta}$$

$$\sec\theta = \frac{1}{\cos\theta} \qquad \cot\theta = \frac{1}{\tan\theta}$$

$$\sin^2\theta + \cos^2\theta = 1 \qquad \tan^2\theta + 1 = \sec^2\theta \qquad \cot^2\theta + 1 = \csc^2\theta$$

Even-Odd Identities

$$\sin(-\theta) = -\sin\theta \qquad \csc(-\theta) = -\csc\theta$$

$$\cos(-\theta) = \cos\theta \qquad \sec(-\theta) = \sec\theta$$

$$\tan(-\theta) = -\tan\theta \qquad \cot(-\theta) = -\cot\theta$$

Double-Angle Formulas

$$\sin 2\theta = 2\sin\theta\cos\theta$$

$$\cos 2\theta = \cos^2\theta - \sin^2\theta$$

$$= 1 - 2\sin^2\theta$$

$$= 2\cos^2\theta - 1$$

$$\tan 2\theta = \frac{2\tan\theta}{1 - \tan^2\theta}$$

Sum and Difference Formulas

$$\sin(\alpha \pm \beta) = \sin\alpha\cos\beta \pm \cos\alpha\sin\beta$$

$$\cos(\alpha \pm \beta) = \cos\alpha\cos\beta \mp \sin\alpha\sin\beta$$

$$\tan(\alpha \pm \beta) = \frac{\tan\alpha \pm \tan\beta}{1 \mp \tan\alpha\tan\beta}$$

Note: The $\mp$ sign means to use the top sign $(-)$ when using the other top signs in the formula. For example, $\cos(\alpha + \beta) = \cos\alpha\cos\beta - \sin\alpha\sin\beta$.

Example

Use the difference formula to find the exact value of $\cos\frac{\pi}{12}$.

Solution

$$\cos\frac{\pi}{12} = \cos\left(\frac{\pi}{4} - \frac{\pi}{6}\right) = \cos\frac{\pi}{4}\cos\frac{\pi}{6} + \sin\frac{\pi}{4}\sin\frac{\pi}{6}$$

$$= \left(\frac{\sqrt{2}}{2}\right)\left(\frac{\sqrt{3}}{2}\right) + \left(\frac{\sqrt{2}}{2}\right)\left(\frac{1}{2}\right)$$

$$= \left(\frac{\sqrt{6}}{4}\right) + \left(\frac{\sqrt{2}}{4}\right)$$

$$= \left(\frac{\sqrt{6} + \sqrt{2}}{4}\right)$$

CHAPTER 6

Exponential and Logarithm Functions

Chapter 6
Exponential and Logarithm Functions

EXPONENTIAL FUNCTIONS

We now consider *exponential functions* of the form $y = a^x$, where $a > 0$ and $a \neq 1$. Notice that this function is different from the power functions $y = x^p$, where the base is the variable x and the exponent is p, a constant. Exponential functions are part of the family of transcendental functions, which were discussed briefly in Chapter 3. The basic graphs of exponential functions are shown in Figure 6.1.

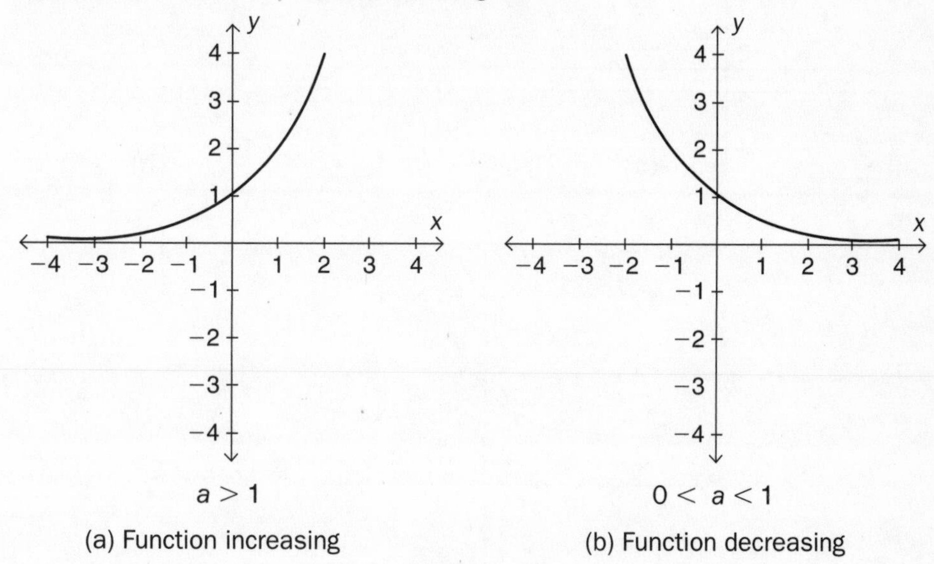

Figure 6.1

Notice from Figure 6.1 that the function is always positive and will always go through the point (0, 1). Notice also in the function $y = a^x$, where $a > 1$, that $y = 0$ (which is the x-axis) is a horizontal asymptote as $x \to -\infty$. For $y = a^x$, where $0 < a < 1$, $y = 0$ is a horizontal asymptote as $x \to \infty$. We exclude $a = 1$ for $y = a^x$ since this is the constant function $y = 1$.

We can use the parent functions of $y = a^x$ for both cases, $a > 1$ and $0 < a < 1$, to sketch translations for functions similar to the parent functions.

Example

Sketch (a) $y = a^x$, (b) $y = a^{x+1}$, and (c) $y = a^x + 1$.

Solution

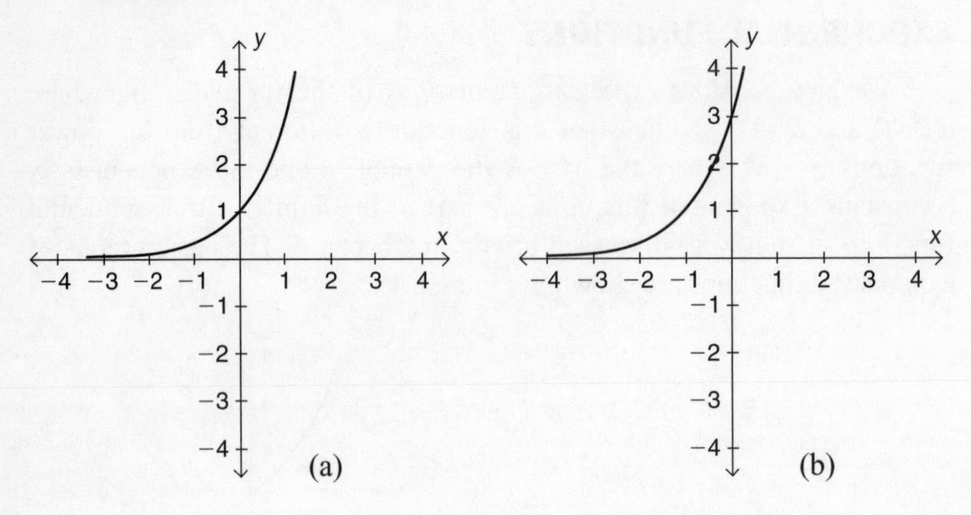

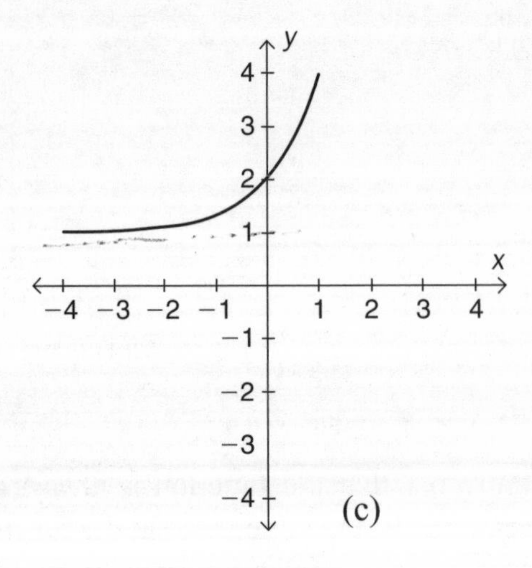

Figure 6.2

If the graph in Figure 6.2(a) is the parent function, then the graph of $y = a^{x+1}$ in Figure 6.2(b) is the parent function multiplied by a, since $y = a^{x+1} = a^x \times a$. For the graph in Figure 6.2(c), the parent function is moved up one unit.

Example

Sketch (a) $y = \left(\frac{1}{a}\right)^x$, (b) $y = \left(\frac{1}{a}\right)^{x+1}$, and (c) $y = \left(\frac{1}{a}\right)^x + 1$.

Solution

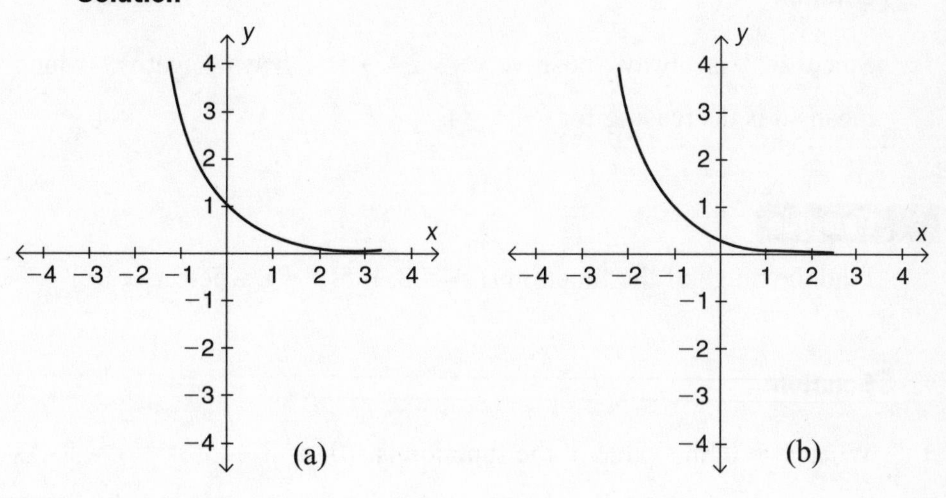

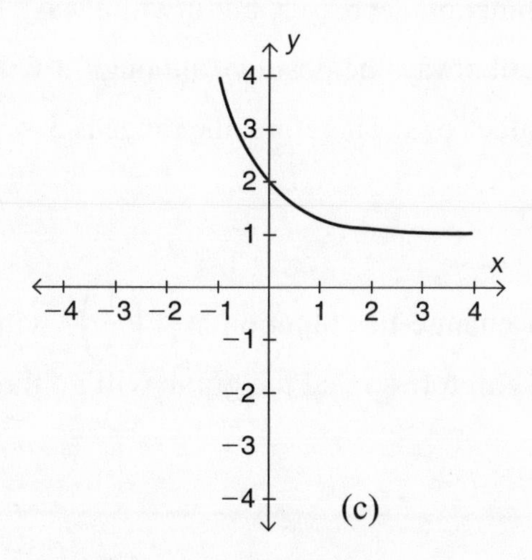

Figure 6.3

Similarly, the graphs in Figure 6.3(b) and (c) are translations of the parent function in Figure 6.3(a). For Figure 6.3(b), the parent function is multiplied by $\frac{1}{a}$ since $y = \left(\frac{1}{a}\right)^{x+1} = \left(\frac{1}{a}\right)^x \times \left(\frac{1}{a}\right)$. The graph in Figure 6.3(c) has moved the parent function up one unit.

Example

For what interval for x is $f(x) = -3(2^{x+1})$ decreasing?

Solution

Since 2^{x+1} is always positive, $-3(2^{x+1})$ is always negative, which means it is decreasing for $(-\infty, \infty)$.

Example

Find the range of the function $f(x) = 3 + 5(3^{-.2x})$, where $x \geq 0$.

Solution

When $x = 0$, the value of the function is $f(0) = 3 + 5(3^{-.2(0)}) = 8$. As x increases, the function decreases, but it will always be greater than 3 because $3^{-.2x}$ will always be positive, although it will be very, very small for large values of x. Therefore, the range is $3 < y \leq 8$.

Example

How would you change the function $y = \left(\frac{1}{3}\right)^x$ without changing the horizontal asymptote so that the graph will go through the point $\left(0, \frac{1}{3}\right)$?

Solution

We first look at the graph of $y = \left(\frac{1}{3}\right)^x$, shown in Figure 6.4.

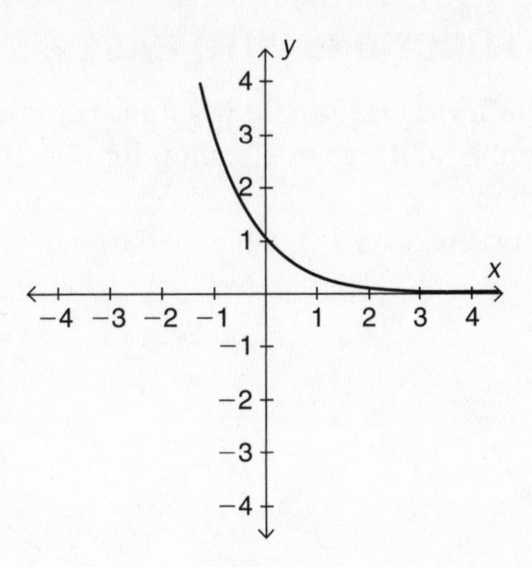

Figure 6.4

We need to change the y-intercept from (0, 1) to $\left(0, \frac{1}{3}\right)$. To do so, we can multiply the function by $\frac{1}{3}$. Therefore, the new function will be $y = \left(\frac{1}{3}\right)^x \times \left(\frac{1}{3}\right)$ or $y = \left(\frac{1}{3}\right)^{x+1}$, and its graph is shown in Figure 6.5.

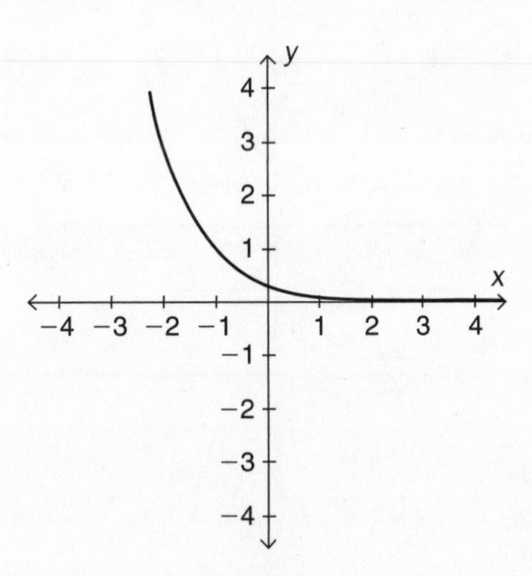

Figure 6.5

Note: If we tried to first translate the parent function, which goes through (0, 1), down $\frac{2}{3}$ units so that it would then go through the point $\left(0, \frac{1}{3}\right)$, we would also change the horizontal asymptote.

EXPONENTIAL FUNCTIONS WITH BASE *e*

Of special significance is the case when the base, which is a constant, is the irrational number e. An approximation for e is 2.718281828…. The sketch of $y = e^x$ should be somewhere between the graphs of $y = 2^x$ and $y = 3^x$ since e is between 2 and 3 (see Figure 6.6).

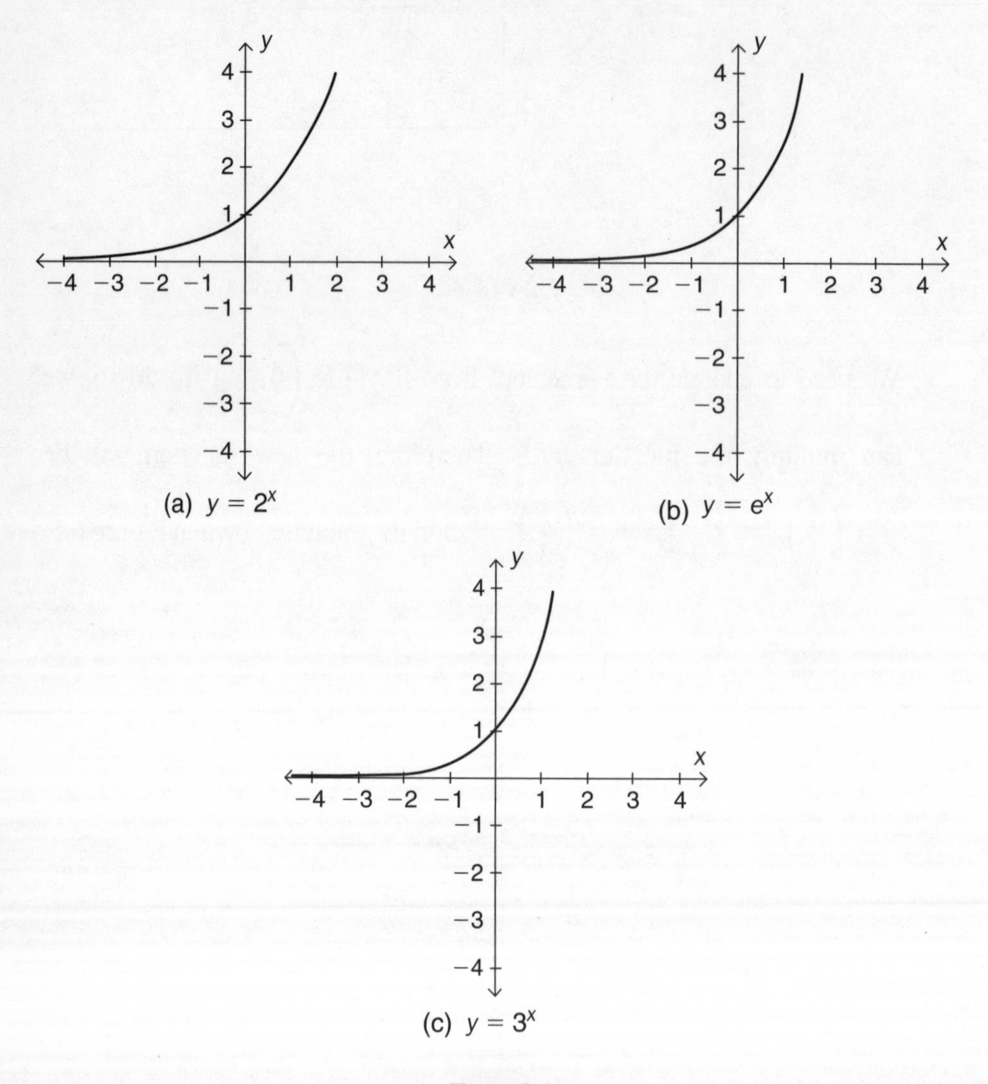

Figure 6.6

The domain for x is all real numbers. The range is all positive y-values. We can use our translations with the parent functions of $y = e^x$ and $y = e^{-x}$ to sketch similar exponential functions with this base.

Example

How would you shift the graph of $y = e^{-x}$ so that $y = 1$ is an asymptote?

Solution

We first graph $y = e^{-x}$, as shown in Figure 6.7. It is the mirror image of $y = e^x$ shown in Figure 6.6(b).

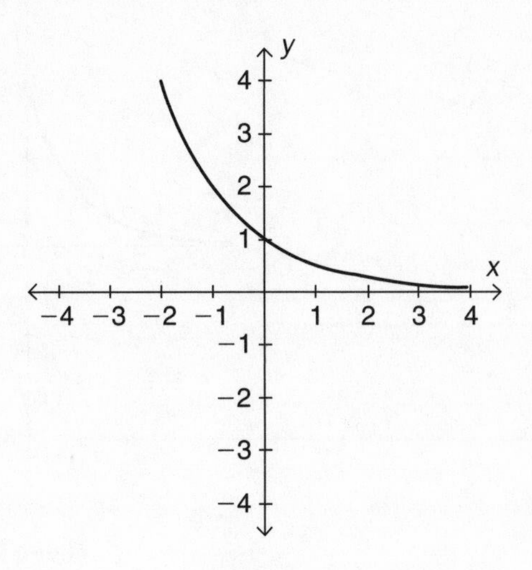

Figure 6.7

For the graph of this function, $y = 0$ is an asymptote. If we want $y = 1$ to be an asymptote, then the graph must be raised one unit. Therefore, the graph of $y = e^{-x} + 1$ (see Figure 6.8) will have a horizontal asymptote at $y = 1$.

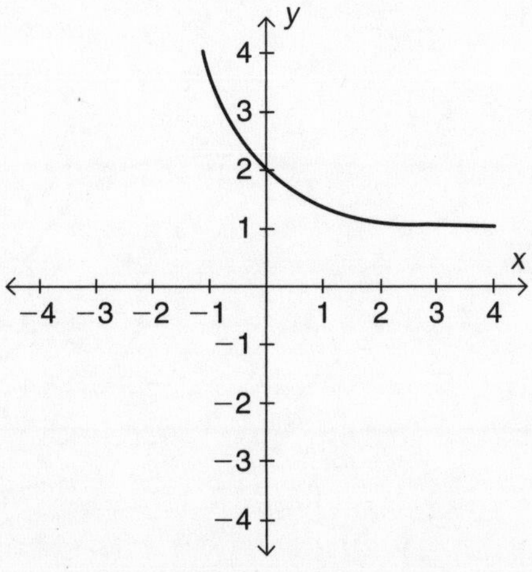

Figure 6.8

Example

Sketch and find the range and domain of the following functions: (a) $y = e^{x+1}$, (b) $y = -e^{x+1}$, and (c) $y = -e^{-(x+1)}$

Solution

(a) $y = e^{x+1}$

Domain: $(-\infty, \infty)$

Range: $(0, \infty)$

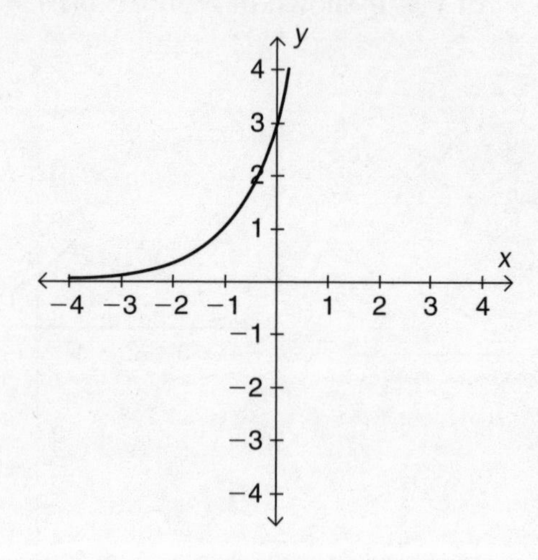

Figure 6.9

(b) $y = -e^{x+1}$

Domain: $(-\infty, \infty)$

Range: $(-\infty, 0)$

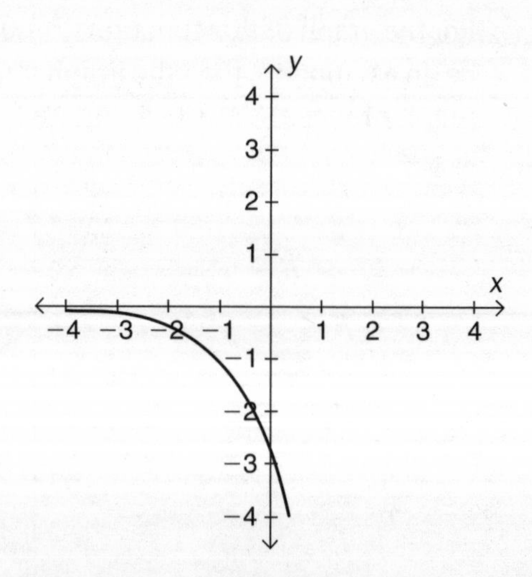

Figure 6.10

(c) $y = -e^{-(x+1)}$

Domain: $(-\infty, \infty)$

Range: $(-\infty, 0)$

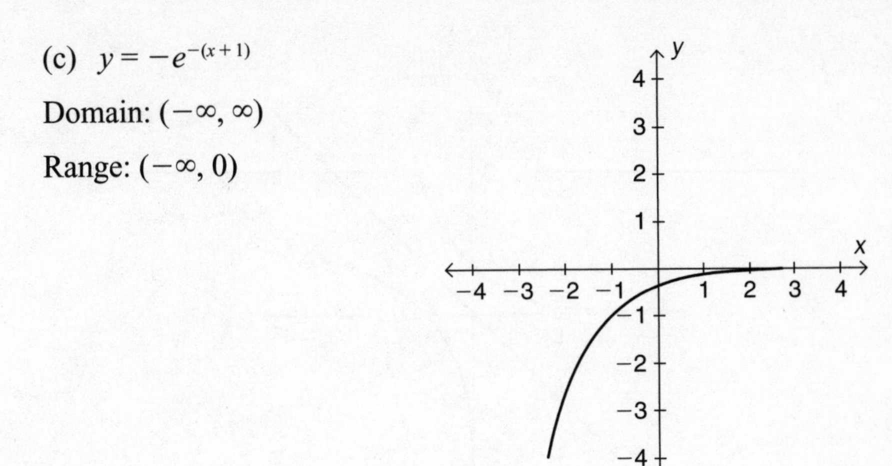

Figure 6.11

INVERSE FUNCTIONS

Recall the graph of $y = a^x$, $a > 1$ (Figure 6.1(a)), reproduced here as Figure 6.12.

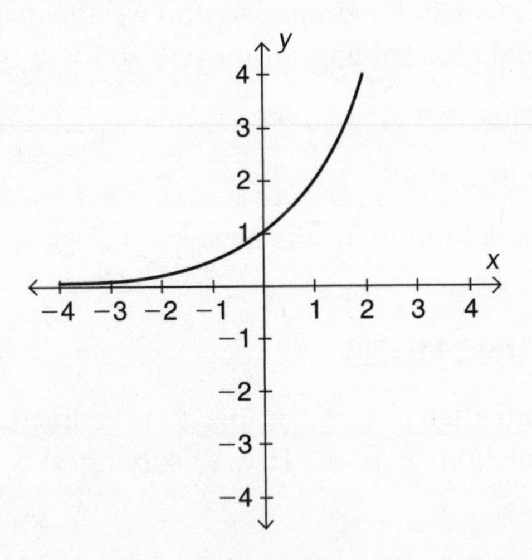

Figure 6.12

Domain: $(-\infty, \infty)$ and Range: $(0, \infty)$

Since the function $y = a^x$, $a > 1$, passes the horizontal line test (and hence is a 1–1 function), the function has an inverse function. We can see the graph of the inverse function by reflecting the original graph about the line $y = x$, as shown in Figure 6.13.

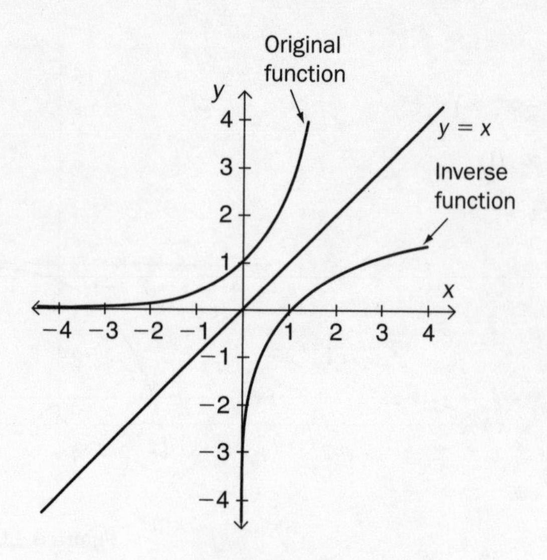

Figure 6.13

Original function	Inverse function
Domain: $(-\infty, \infty)$	Domain: $(0, \infty)$
Range: $(0, \infty)$	Range: $(-\infty, \infty)$

To find the inverse function, we follow the two-step process of replacing x and y and then solving for the new y.

Original function: $y = a^x$

Inverse function: $x = a^y$

We now give a definition for the new y.

LOGARITHM FUNCTIONS

The logarithmic function, $y = \log_a x$, is defined as the inverse of the exponential function, $y = a^x$. Hence, solving $x = a^y$ for y gives us $y = \log_a x$.

For the function $y = \log_a x$, we can find the domain, range, and then sketch. (Figure 6.14 is taken from Figure 6.13.)

Domain: $(0, \infty)$

Range: $(-\infty, \infty)$

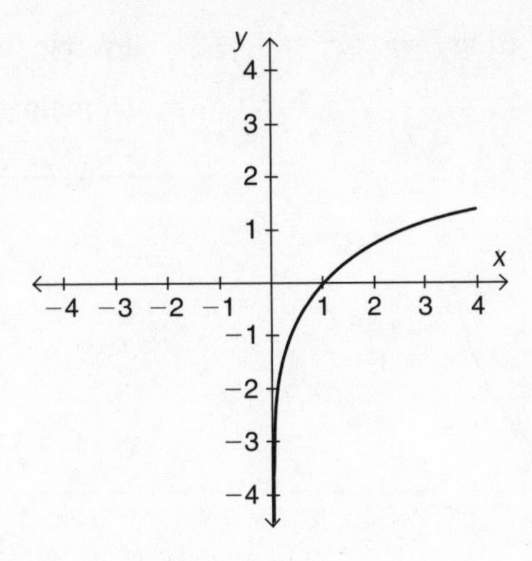

Figure 6.14

Common Logarithms

The inverse of $y = 10^x$ is a special logarithm, $y = \log x$, and is called the *common logarithm* function. The base in $y = \log x$ is understood to be 10. Since $y = 10^x$ and $y = \log x$ are inverse functions, one process will "undo" the other. Therefore,

$\log(10^x) = x$ for all x

$10^{\log x} = x$ for $x > 0$

Note that $\log 1 = 0$ (since $10^0 = 1$) and $\log 10 = 1$ (since $10^1 = 10$).

Natural Logarithms

The inverse of $y = e^x$ is also a special logarithm function, called the *natural logarithm* function, and is designated by $y = \ln x$.

Given $y = e^x$, the inverse function is $x = e^y$, or solving for y, $y = \ln x$.

Original function $f(x) = e^x$

Domain: $(-\infty, \infty)$

Range: $(0, \infty)$

Inverse function $f^{-1}(x) = \ln x$

Domain: $(0, \infty)$

Range: $(-\infty, \infty)$

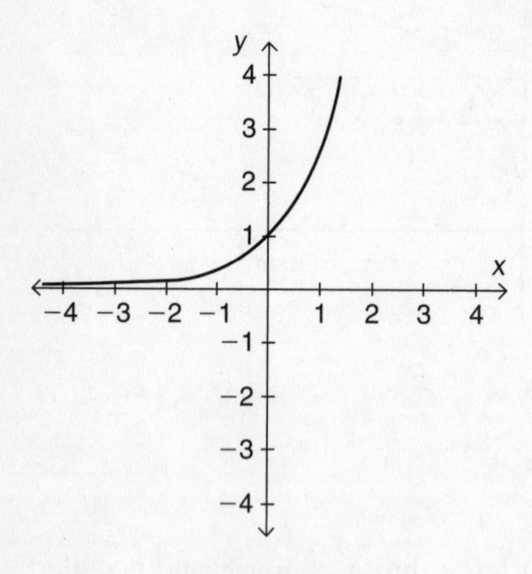

Figure 6.15

Figure 6.16

Because these are inverse functions,

$\ln(e^x) = x$ for all x

$e^{\ln x} = x$ for $x > 0$

Note that $\ln 1 = 0$ (since $e^0 = 1$) and $\ln e = 1$ (since $e^1 = e$).

Example

Sketch $y = 2 - \ln x$ by using the graph of the parent function $y = \ln x$.

Solution

We begin with the graph of the parent function $y = \ln x$ (Figure 6.16).

We now graph $y = -\ln x$, shown in Figure 6.17.

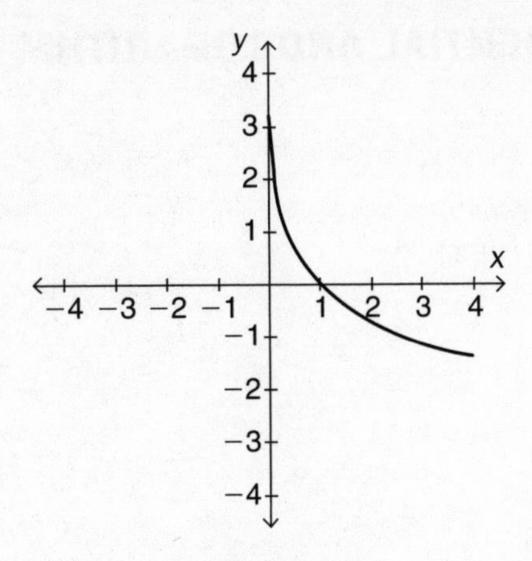

Figure 6.17

Finally, we add 2 to the function, which raises the graph by two units, as shown in Figure 6.18. $y = 2 - \ln x$

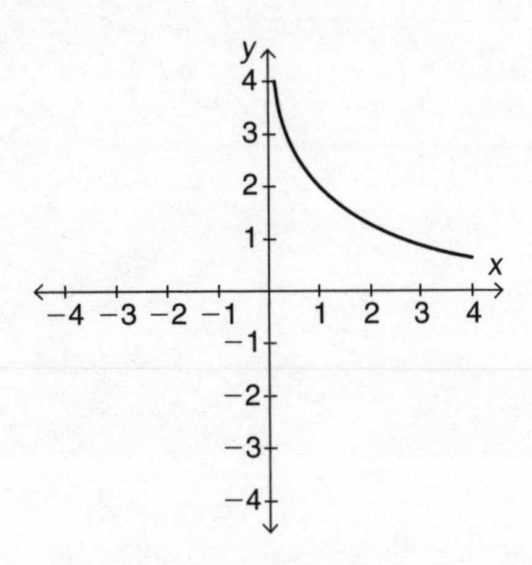

Figure 6.18

Quick Review of the Rules of Logarithms

We see that the rules for logarithms are analogous to the laws of exponents:

$$x^a \times x^b = x^{a+b}; \frac{x^a}{x^b} = x^{a-b}; (x^a)^b = x^{ab}$$

$\log(a \times b) = \log a + \log b$ $\qquad$ $\ln(a \times b) = \ln a + \ln b$

$\log\left(\frac{a}{b}\right) = \log a - \log b$ $\qquad$ $\ln\left(\frac{a}{b}\right) = \ln a - \ln b$

$\log(a^r) = r \log a$ $\qquad$ $\ln(a^r) = r \ln a$

SOLVING EXPONENTIAL AND LOGARITHM EQUATIONS

Example

Solve $P = P_o e^{3x}$ for x.

Solution

$P = P_o e^{3x}$

We first divide by P_o.

$$\frac{P}{P_o} = e^{3x}$$

We now rewrite using logarithms and then use the rule $\ln(e^u) = u$ for all u.

$$\ln\left(\frac{P}{P_o}\right) = \ln(e^{3x})$$

$$\ln\left(\frac{P}{P_o}\right) = 3x$$

$$\frac{1}{3}\ln\left(\frac{P}{P_o}\right) = x$$

Example

Solve $e^x - 6e^{-x} + 1 = 0$.

Solution

To solve, we first multiply everything on both sides by e^x to get rid of the negative exponent. Thus, we have

$$e^x(e^x - 6e^{-x} + 1) = e^x(0)$$

$$e^{2x} - 6 + e^x = 0$$

Rearranging the terms, we get

$$e^{2x} + e^x - 6 = 0$$

This is in a form similar to a quadratic equation. We now factor to solve.

$(e^x - 2)(e^x + 3) = 0$

Setting each factor equal to 0 and solving gives us:

$e^x = 2$ and $e^x = -3$ (which is impossible since e^x is always positive)

To solve $e^x = 2$, we rewrite the equation by taking the natural logarithm of both sides.

$\ln(e^x) = \ln 2$

Since $\ln(e^x) = x$, we get

$x = \ln 2$

We now check this solution in our original equation.

$e^{\ln 2} - 6e^{-\ln 2} + 1 ? 0$

Since $e^{\ln 2} = 2$ and $e^{-\ln 2} = \frac{1}{e^{\ln 2}} = \frac{1}{2}$, we have

$2 - 6\left(\frac{1}{2}\right) + 1 = 0$, which checks our solution of $x = \ln 2$.

An alternate solution for $e^x - 6e^{-x} + 1 = 0$ is to let $y = e^x$. Then the equation can be written as $y - \frac{6}{y} + 1 = 0$, or multiplying through by y we get $y^2 - 6 + y = 0$. We can rearrange the terms, factor, set each factor equal to 0, and solve.

$y^2 + y - 6 = 0$

$(y - 2)(y + 3) = 0$

$y = 2 \quad y = -3$ (which we discard since $y = e^x \neq -3$)

We now substitute back and solve for x, as above.

$e^x = 2$

$x = \ln 2$

Example

Solve $4^{x^2-3x} = 1$ for x.

Solution

To solve, we write the equation using base 4 on each side.

$4^{x^2-3x} = 4^0$

Now we equate the exponents.

$x^2 - 3x = 0$

We factor, set each factor equal to 0, and solve.

$x(x - 3) = 0$

$x = 0$ and $x - 3 = 0$

$x = 3$

The two solutions $x = 0$ and $x = 3$ both check in the equation.

Example

Solve $\log_2 64 = 5x$ for x.

Solution

We solve for x by rewriting the equation in an exponential form.

$\log_2 64 = 5x$ can be rewritten as $2^{5x} = 64$.

To solve the exponential equation, we write both sides with base 2.

$2^{5x} = 2^6$

Now we can equate the exponents and solve.

$5x = 6$

$x = \frac{6}{5}$, which checks in the equation.

Example

Given $f(x) = ab^x$ and $g(x) = ax^b$, where a and b are constants and $b > 0$. If $f(1) = 6$ and $g(1) = 7$, find a and b.

Solution

$f(1) = ab^1 = ab$, but $f(1) = 6$, so $ab = 6$

$g(1) = a(1)^b = a$, but $g(1) = 7$, so $a = 7$

Since $ab = 6$ and $a = 7$, we get $b = \frac{6}{7}$.

Example

Given f and g as defined in the previous example, find the composition functions $f \circ g$ and $g \circ f$.

Solution

$f \circ g = f(g(x)) = f(ax^b) = ab^{(ax^b)}$

$g \circ f = g(f(x)) = g(ab^x) = a(ab^x)^b = a(a^b b^{bx}) = a^{b+1}b^{bx}$

GROWTH AND DECAY PROBLEMS

Exponential functions provide a model for how things grow or decay as a function of time. These functions are generally represented by $y = p_o a^t$, where $a > 0$, p_o is the initial value, and t is time. If $a > 1$, there is exponential growth. If $0 < a < 1$, there is exponential decay.

The corresponding exponential growth (or decay) function for base e is $P = P_o e^{kt}$, where P is an ending amount, P_o is an initial amount, k is a continuous growth (or decay) rate, and t is time. If $k > 0$, there is exponential growth, and if $k < 0$, there is exponential decay.

Example

You start a new job at a base salary of $35,000, with a guaranteed salary increase of 3% per year. How many years before your base salary increases to $100,000?

Solution

The function that will describe the growth is $y = p_o a^t$, where y is the ending salary and p_o is the beginning salary.

$$100{,}000 = 35{,}000(1.03)^t$$

$$\frac{100000}{35000} = (1.03)^t$$

Reducing, we obtain

$$\frac{20}{7} = (1.03)^t$$

We solve for t by writing the equation with logarithms.

$$\ln\left(\frac{20}{7}\right) = \ln(1.03)^t$$

$$\ln\left(\frac{20}{7}\right) = t \times \ln(1.03)$$

$$t = \frac{\ln\left(\frac{20}{7}\right)}{\ln(1.03)} \approx 35.52 \text{ years}$$

Note: The values for $\ln\left(\frac{20}{7}\right)$ and $\ln(1.03)$ are obtained from a calculator or table of natural logs.

Example

The population, in thousands, of City A can be modeled by $y_1 = 48(1.028)^t$, and the population, in thousands, of City B can be modeled by $y_2 = 35(1.049)^t$, where t in both models is the number of years since 2002. When will the population of the two cities be equal, using these growth models?

Solution

We want to solve for the value of t that will make the populations equal.

$$48(1.028)^t = 35(1.049)^t$$

$$\frac{48}{35} = \frac{(1.049)^t}{(1.028)^t} = \left(\frac{1.049}{1.028}\right)^t$$

$$\ln\left(\frac{48}{35}\right) = \ln\left(\frac{1.049}{1.028}\right)^t$$

$$\ln\left(\frac{48}{35}\right) = t \times \ln\left(\frac{1.049}{1.028}\right)$$

$$t = \frac{\ln\left(\frac{48}{35}\right)}{\ln\left(\frac{1.049}{1.028}\right)} \approx 15.63 \text{ years}$$

Example

A substance decays exponentially. If initially there was 200 mg of the substance and after 15 hours there was only 120 mg, how long will it take for the initial substance to decay to 5 mg?

Solution

We first must find the value of k. We use the exponential decay formula, $P = P_o e^{kt}$, where P (the ending amount) is 120 mg, P_o (the initial amount) is 200 mg, and t is 15 hours.

$$P = P_o e^{kt}$$

$$120 = 200e^{k(15)}$$

$$\frac{120}{200} = e^{15k}$$

Reducing and rewriting with logarithms, we have

$$\ln\left(\frac{3}{5}\right) = \ln\left(e^{15k}\right)$$

Since $\ln(e^x) = x$, we have

$$\ln\left(\frac{3}{5}\right) = 15k$$

$$k = \frac{\ln\left(\frac{3}{5}\right)}{15} \approx -.034$$

We now use this value of k to solve for the time it will take the initial amount (200 mg) to decay to 5 mg.

$$5 = 200e^{(-.034)t}$$

$$\frac{5}{200} = e^{(-.034)t}$$

Reducing and rewriting with logarithms, we will have

$$\ln\left(\frac{1}{40}\right) = \ln\left(e^{-.034t}\right)$$

$$\ln\left(\frac{1}{40}\right) = -.034t$$

$$t = \frac{\ln\left(\frac{1}{40}\right)}{-.034} \approx 108.5 \text{ hours}$$

Example

A certain strain of bacteria, Strain A, doubles every 12 hours. Strain B grows by 2% per hour. After 20 hours, which strain has the largest population?

Solution

We assume they begin with the same initial amount of N_0 at time 0.

Strain A — $N = N_0 2^{\frac{t}{12}}$

Strain B — $N = N_0(1.02)^t$

After 20 hours:

$N = N_0 2^{\frac{20}{12}} \approx N_0(3.1748)$ $\quad$ $N = N_0(1.02)^{20} \approx N_0(1.4859)$

There will be more than twice the number of Strain A bacteria than Strain B bacteria after 20 hours.

SPECIAL FUNCTIONS

There are certain simple expressions involving exponential functions that occur frequently in applications of mathematics, especially in physics and engineering. These are called hyperbolic functions.

The hyperbolic functions are related to the graph of $x^2 - y^2 = 1$, called a hyperbola (see Chapter 7), in the same way as trigonometric functions are related to the graph of $x^2 + y^2 = 1$, which is a circle. Two hyperbolic functions, $\sin h$ (hyperbolic sine) and $\cos h$ (hyperbolic cosine), are defined as follows:

$$\sin h\, x = \frac{e^x - e^{-x}}{2} \qquad \text{and} \qquad \cos h\, x = \frac{e^x + e^{-x}}{2}$$

Example

Prove $\cos h^2 x - \sin h^2 x = 1$.

Solution

$$\cos h^2\, x - \sin h^2\, x = \left(\frac{e^x + e^{-x}}{2}\right)^2 - \left(\frac{e^x - e^{-x}}{2}\right)^2$$

$$= \left(\frac{e^{2x} + 2 + e^{-2x}}{4}\right) - \left(\frac{e^{2x} - 2 + e^{-2x}}{4}\right)$$

$$= \frac{4}{4} = 1$$

CHAPTER 7
Conics

Chapter 7
Conics

Conic equations have the general form

$$Ax^2 + Bxy + Cy^2 + Ex + Fy + G = 0$$

where the coefficients are constants, and A, B, and C cannot all equal 0. They are formed by the intersection of a right circular cone and a plane. It is important to recognize the conic from its equation and to be able to gather enough information to sketch it. To do so, we represent the conic by one of the following standard forms.

$(x-h)^2 + (y-k)^2 = r^2$	Circle, center (h, k) and radius r
$\frac{(x-h)^2}{a^2} + \frac{(y-k)^2}{b^2} = 1$	Ellipse, center (h, k), major axis parallel to x-axis, $a^2 > b^2$
$\frac{(x-h)^2}{b^2} + \frac{(y-k)^2}{a^2} = 1$	Ellipse, center (h, k), major axis parallel to y-axis, $a^2 > b^2$
$\frac{(x-h)^2}{a^2} - \frac{(y-k)^2}{b^2} = 1$	Hyperbola, center (h, k), opens right/left
$\frac{(y-k)^2}{a^2} - \frac{(x-h)^2}{b^2} = 1$	Hyperbola, center (h, k), opens up/down
$y = a(x-h)^2 + k$	Parabola, vertex (h, k), opens up or down
$x = a(y-k)^2 + h$	Parabola, vertex (h, k), opens right or left

Also, there are special cases in which the intersection of the cone and plane is a point, a line, or a pair of intersecting lines. These are called degenerate conics.

Note: In this chapter, we assume $B = 0$ in the general form of the conic equation. See the end of this chapter for a brief discussion of how B affects the rotation of the conics.

Another way to determine the type of conic is to examine $B^2 - 4AC$. If $B^2 - 4AC > 0$, the conic is a hyperbola or the degenerate hyperbola of two intersecting lines; if $B^2 - 4AC = 0$, the conic is a parabola or the degenerate case, which is a line; and if $B^2 - 4AC < 0$, the conic is an ellipse or a circle, or the degenerate case, which is a point.

CIRCLE

If $A = C$ in the general form $Ax^2 + Bxy + Cy^2 + Ex + Fy + G = 0$, the conic is a circle. By completing the square, we can rewrite the equation in its standard form. (See Chapter 2, Algebra Review, for a more detailed review of the process of completing the square.)

Example

Identify the conic and sketch.

$x^2 + y^2 + 6x + 4y - 3 = 0$

Solution

Since $A = C$, the equation represents a circle. We first complete the square to place the equation in standard form.

$(x^2 + 6x + \_\_\_) + (y^2 + 4y + \_\_\_) = 3 + \_\_\_ + \_\_\_$

$(x^2 + 6x + \underline{\ 9\ }) + (y^2 + 4y + \underline{\ 4\ }) = 3 + \underline{\ 9\ } + \underline{\ 4\ }$

$$(x + 3)^2 + (y + 2)^2 = 16$$

This is a circle with center $(-3, -2)$ and radius 4. The sketch is shown in Figure 7.1.

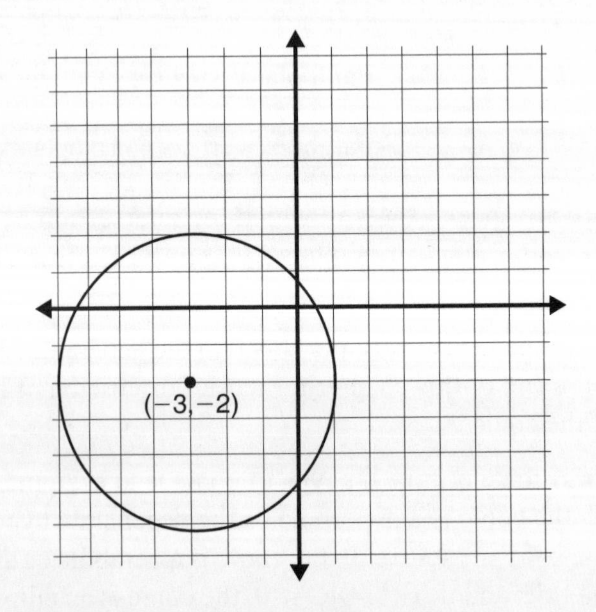

Figure 7.1

Example

Identify the conic and sketch.

$3x^2 + 3y^2 - 6x + 1 = 0$

Solution

Since $A = C$, the equation represents a circle. We complete the square to place the equation into standard form. First, we divide everything by 3 so that the coefficient of both x^2 and y^2 is 1.

$$x^2 + y^2 - 2x + \frac{1}{3} = 0$$

$$(x^2 - 2x + \underline{\quad}) + y^2 = -\frac{1}{3} + \underline{\quad}$$

$$(x^2 - 2x + \underline{1}) + y^2 = -\frac{1}{3} + \underline{1}$$

$$(x - 1)^2 + y^2 = \frac{2}{3}$$

The circle has center (1, 0) and radius $\sqrt{\frac{2}{3}}$ or $\frac{\sqrt{6}}{3}$. The sketch is shown in Figure 7.2.

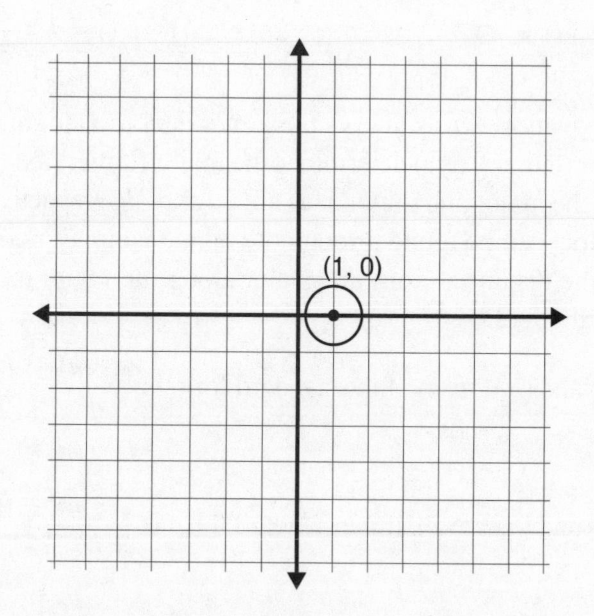

Figure 7.2

ELLIPSE

If $A \neq C$ and A, C have the same signs in the general form, $Ax^2 + Bxy + Cy^2 + Ex + Fy + G = 0$, the conic is an ellipse. By definition, an ellipse is the set of all points for which the sum of the distances from two fixed points (called foci) is constant (Figure 7.3). The standard form is either

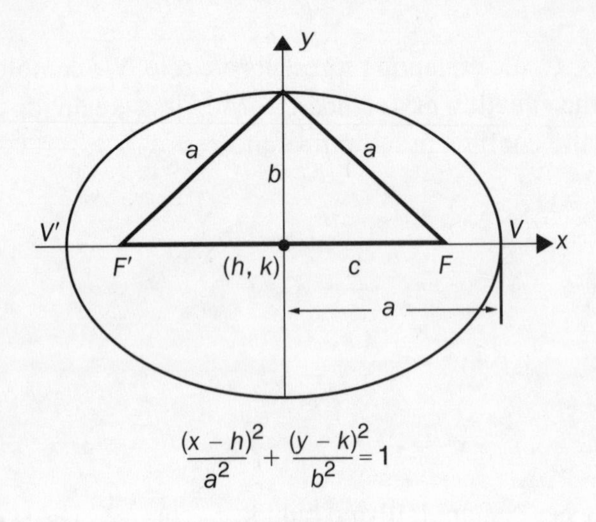

Figure 7.3

$$\frac{(x-h)^2}{a^2} + \frac{(y-k)^2}{b^2} = 1 \quad \text{or} \quad \frac{(x-h)^2}{b^2} + \frac{(y-k)^2}{a^2} = 1$$

In these equations, a^2 is always the larger denominator and the variable in its numerator (either x or y) determines the major (larger) axis of the ellipse. The lengths of the major and minor axes are $2a$ and $2b$, respectively. The coordinates of the foci can be found through the relationship: $c^2 = a^2 - b^2$, where c and $-c$ are the distances from the center along the major axis. Hence, the coordinates of the foci are:

$(h + c, k)$ and $(h - c, k)$, if the standard form is $\frac{(x-h)^2}{a^2} + \frac{(y-k)^2}{b^2} = 1$,

or

$(h, k + c)$ and $(h, k - c)$, if the standard form is $\frac{(x-h)^2}{b^2} + \frac{(y-k)^2}{a^2} = 1$

Example

Identify the conic, determine the coordinates of the center and foci, and then sketch the following conic equation:

$9x^2 + 4y^2 + 36x - 24y + 36 = 0$

Solution

Since $A \neq C$ and A and C have the same signs, we see the conic is an ellipse. To identify the center and foci, we need to complete the square and put the equation into standard form. We factor 9 from the coefficients of x^2 and x, and 4 from the coefficients of y^2 and y, and then complete the squares.

$$9(x^2 + 4x) + 4(y^2 - 6y) = -36$$

$$9(x^2 + 4x + \underline{\quad}) + 4(y^2 - 6y + \underline{\quad}) = -36 + \underline{\quad} + \underline{\quad}$$

$$9(x^2 + 4x + \underline{\ 4\ }) + 4(y^2 - 6y + \underline{\ 9\ }) = -36 + \underline{\ 36\ } + \underline{\ 36\ }$$

$$9(x + 2)^2 + 4(y - 3)^2 = 36$$

Since the standard form for an ellipse is equal to 1, we must divide everything by 36. After reducing the fractions, we get . . .

$$\frac{(x+2)^2}{4} + \frac{(y-3)^2}{9} = 1$$

. . .from which we get the following values:

Center: $(h, k) = (-2, 3)$

$a^2 = 9 \Rightarrow a = \pm 3$

$b^2 = 4 \Rightarrow b = \pm 2$

$c^2 = a^2 - b^2 = 9 - 4 = 5$, or $c = \pm\sqrt{5}$

Since a^2 (the larger number) is the denominator of the y-term in this example, the major axis is parallel to the y-axis and the minor axis is parallel to the x-axis. The distance along the major axis is $2a$ (or 6) and the distance along the minor axis is $2b$ (or 4).

To sketch, we locate the center and move a units above and below the center to get vertices (endpoints of the major axis) of the ellipse and b units right and left to get the co-vertices (endpoints of the minor axis). The coordinates of the foci are $(h, k + c)$ and $(h, k - c)$, or $F2(-2, 3 + \sqrt{5})$ and $F1(-2, 3 - \sqrt{5})$. The sketch is shown in Figure 7.4.

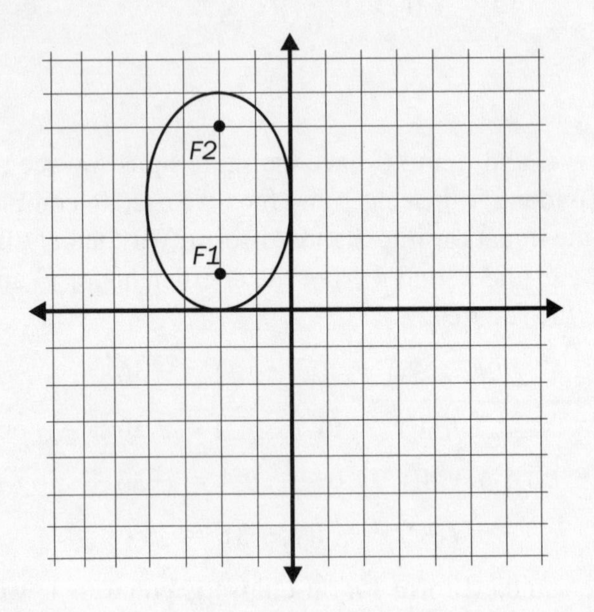

Figure 7.4

Note: We move a units above and below the center of this ellipse since the major axis is parallel to the y-axis. We move b units right and left since the minor axis is parallel to the x-axis.

Example

Identify the coordinates of the center, the foci, the vertices along the major axis, and the co-vertices along the minor axis of the conic equation

$12x^2 + 20y^2 - 12x + 40y - 37 = 0$

Solution

The conic is an ellipse since $A \neq C$ and A and C have the same signs. To identify the center and foci, we need to complete the square and put the equation into standard form. We factor 12 from the coefficients of x^2 and x, and 20 from the coefficients of y^2 and y.

$$12(x^2 - x) + 20(y^2 + 2y) = 37$$

$$12(x^2 - x + \_\_\_) + 20(y^2 + 2y + \_\_\_) = 37 + \_\_\_ + \_\_\_$$

$$12\left(x^2 - x + \underline{\frac{1}{4}}\right) + 20(y^2 + 2y + \underline{1}) = 37 + \underline{3} + \underline{20}$$

$$12\left(x - \frac{1}{2}\right)^2 + 20(y + 1)^2 = 60$$

Since the standard form for an ellipse is equal to 1, we divide everything by 60 and reduce the fractions to get

$$\frac{\left(x - \frac{1}{2}\right)^2}{5} + \frac{(y + 1)^2}{3} = 1$$

Center: $(h, k) = \left(\frac{1}{2}, -1\right)$

$a^2 = 5 \Rightarrow a = \pm\sqrt{5}$

$b^2 = 3 \Rightarrow b = \pm\sqrt{3}$

$c^2 = a^2 - b^2 = 5 - 3 = 2$, or $c = \pm\sqrt{2}$

Since a^2 (the larger number) is the denominator of the x-term in this problem, the major axis is parallel to the x-axis and the minor axis is parallel to the y-axis. The coordinates of the vertices are $(h + a, k)$ and $(h - a, k)$, or $\left(\frac{1}{2}+\sqrt{5},-1\right)$ and $\left(\frac{1}{2} - \sqrt{5}, -1\right)$. The co-vertices are $(h, k + b)$ and $(h, k - b)$, or $\left(\frac{1}{2},-1+\sqrt{3}\right)$ and $\left(\frac{1}{2}, -1 - \sqrt{3}\right)$. Coordinates of the foci are $(h + c, k)$ and $(h - c, k)$ or $\left(\frac{1}{2} + \sqrt{2}, -1\right)$ $\left(\frac{1}{2} - \sqrt{2}, -1\right)$. The sketch is shown in Figure 7.5.

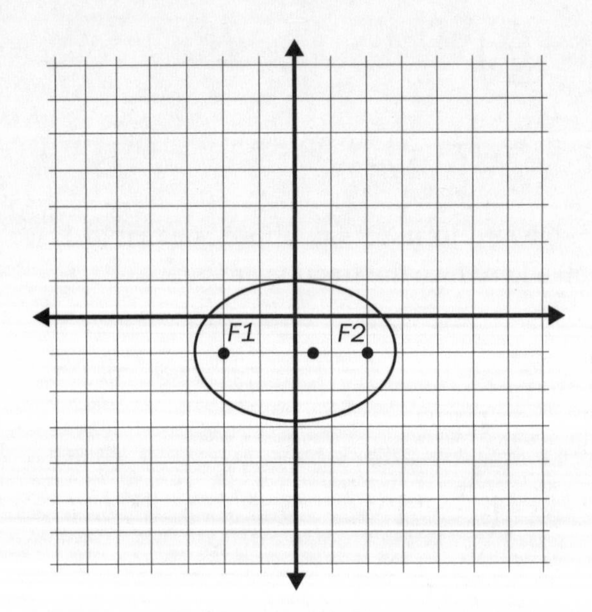

Figure 7.5

HYPERBOLA

If A and C have different signs in the general form $Ax^2 + Bxy + Cy^2 + Ex + Fy + G = 0$, the conic is a hyperbola (Figure 7.6).

By definition, a hyperbola is the set of all points for which the differences of the distances from two fixed points (called foci) is constant. A hyperbola is actually two disconnected curves, each going around a focus point. The standard form is either

$$\frac{(x-h)^2}{a^2} - \frac{(y-k)^2}{b^2} = 1 \quad \text{or} \quad \frac{(y-k)^2}{a^2} - \frac{(x-h)^2}{b^2} = 1$$

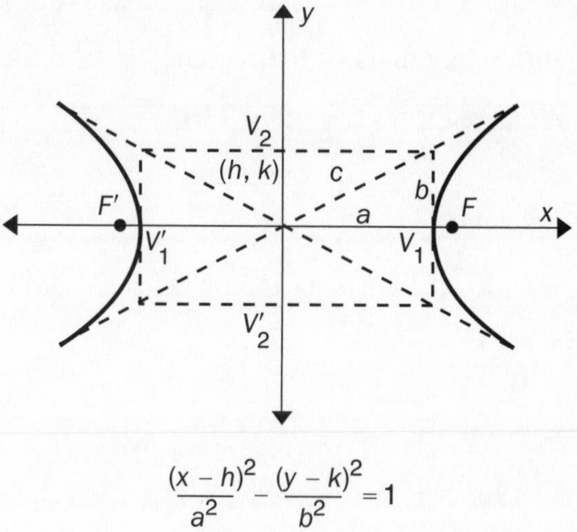

$$\frac{(x-h)^2}{a^2} - \frac{(y-k)^2}{b^2} = 1$$

Figure 7.6

The center of the hyperbola in each case is (h, k). The axis containing the center, the vertices, and the foci is called the transverse axis. In the first equation above, the transverse axis is parallel to the x-axis. The coordinates of the vertices are $V_1(h + a, k)$ and $V'_1(h - a, k)$ for $\frac{(x-h)^2}{a^2} - \frac{(y-k)^2}{b^2} = 1$, and the coordinates of the foci are $(h + c, k)$ and $(h - c, k)$. There are "pseudo-vertices" that help with sketching, but are *not* points on the hyperbola, at $V_2(h, k + b)$ and $V'_2(h, k - b)$. These are along the conjugate axis, which is the axis perpendicular to the transverse axis through the center of the hyperbola. For the second equation above, $\frac{(y-k)^2}{a^2} - \frac{(x-h)^2}{b^2} = 1$, the transverse axis is parallel to the y-axis, the coordinates of the vertices are $(h, k + a)$ and $(h, k - a)$, the coordinates of the foci are $(h, k + c)$ and $(h, k - c)$, and the coordinates of the pseudo-vertices are $(h + b, k)$ and $(h - b, k)$.

Summarizing, for the equation $\frac{(x-h)^2}{a^2} - \frac{(y-k)^2}{b^2} = 1$, we can locate two vertices on the hyperbola by going a units right and left from the center. We can locate the pseudo-vertices by going b units up and down from the center. For the equation $\frac{(y-k)^2}{a^2} - \frac{(x-h)^2}{b^2} = 1$, we locate the two vertices by going a units up and down from the center and the two pseudo-vertices by going b units right and left of the center. Using these four points we sketch a rectangle so that the diagonals of the rectangles form the asymptotes for the hyperbola (see dashed lines in Figure 7.6).

Example

Sketch and identify the vertices and pseudo-vertices of

$$(y+1)^2 - \frac{(x-2)^2}{9} = 1$$

Solution

We rewrite the equation to identify a^2.

$$\frac{(y+1)^2}{1} - \frac{(x-2)^2}{9} = 1$$

Center: $(h, k) = (2, -1)$

$a^2 = 1$, or $a = \pm 1$

$b^2 = 9$, or $b = \pm 3$

The vertices can be found by moving one unit up and down from the center. Their coordinates are $(h, k + a)$ and $(h, k - a)$, or $(2, 0)$ and $(2, -2)$. The pseudo-vertices are at $(h + b, k)$ and $(h - b, k)$, or $(5, -1)$ and $(-1, -1)$. We draw the rectangle from these four points, so that the diagonals, which pass through the center, become the asymptotes of the hyperbola, as shown in Figure 7.7.

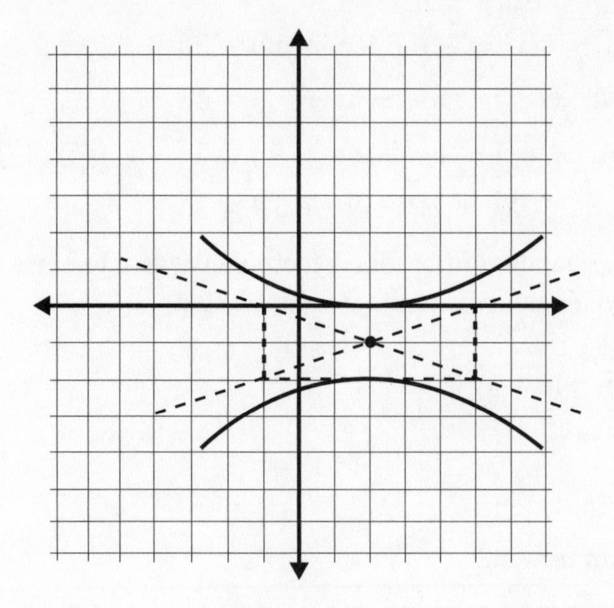

Figure 7.7

The coordinates of the foci can be found through the relationship: $c^2 = a^2 + b^2$, where c and $-c$ are the distances from the center along the major axis. Hence, the coordinates of the foci are:

$(h + c, k)$ and $(h - c, k)$, if the standard form is $\dfrac{(x-h)^2}{a^2} - \dfrac{(y-k)^2}{b^2} = 1$

or

$(h, k + c)$ and $(h, k - c)$, if the standard form is $\dfrac{(y-k)^2}{a^2} + \dfrac{(x-h)^2}{b^2} = 1$

Example

Identify the conic, the coordinates of the center, the vertices, and the foci, and then sketch the following conic equation:

$9x^2 - 4y^2 + 36x - 24y - 36 = 0$

Solution

Since A and C have different signs, the conic is a hyperbola. To identify the center, vertices, and foci, we need to complete the square and put the equation into standard form. We factor 9 from the coefficients of x^2 and x, and -4 from the coefficients of y^2 and y.

$$9(x^2 + 4x) - 4(y^2 + 6y) = 36$$

$$9(x^2 + 4x + \_\_) - 4(y^2 + 6y + \_\_) = 36 + \_\_ + \_\_$$

$$9(x^2 + 4x + \underline{4}) - 4(y^2 + 6y + \underline{9}) = 36 + \underline{(-36)} + \underline{36}$$

$$9(x + 2)^2 - 4(y + 3)^2 = 36$$

Since the standard form for a hyperbola is equal to 1, we divide everything by 36, and reduce the fractions to get

$$\frac{(x+2)^2}{4} - \frac{(y+3)^2}{9} = 1$$

Center: $(h, k) = (-2, -3)$

$a^2 = 4$, or $a = \pm 2$

$b^2 = 9$, or $b = \pm 3$

$c^2 = a^2 + b^2 = 4 + 9 = 13$, or $c = \pm\sqrt{13}$

Coordinates of vertices: $(h + a, k)$ and $(h - a, k)$, or $(0, -3)$ and $(-4, -3)$

Since the x-term is positive, the transverse axis is parallel to the x-axis and the conjugate axis, containing the pseudo-vertices $(h, k + b)$ and $(h, k - b)$, or $(-2, 0)$ and $(-2, -6)$, is parallel to the y-axis.

To sketch the hyperbola, we locate the center and move a (or 2) units right and left of the center to get the vertices of the hyperbola and b (or 3) units up and down to get the pseudo-vertices. The coordinates of the foci are $(h + c, k)$ and $(h - c, k)$, or $F2(-2 + \sqrt{13}, -3)$ and $F1(-2 - \sqrt{13}, -3)$.

Note: We move a units right and left since the transverse axis is parallel to the x-axis. We move b units up and down since the conjugate axis is parallel to the y-axis.

We can determine one of the equations of the asymptotes by writing the equation of the line through the center (h, k) and one of the corners of the rectangle, say $(h + a, k + b)$. We first determine the slope of the line: $m = \dfrac{(k+b)-k}{(h+a)-h} = \dfrac{b}{a}$. Now we can use the slope and one of the points, say the center (h, k), and write the equation of the line.

$$y - k = \frac{b}{a}(x - h), \quad \text{or} \quad y = \frac{b}{a}x + \left(k - \frac{b}{a}h\right)$$

We can determine the equation of the other asymptote in a similar manner by using the center and an adjacent corner, say $(h - a, k + b)$. This equation is

$$y - k = -\frac{b}{a}(x - h) \qquad \text{or} \qquad y = -\frac{b}{a}x + \left(k + \frac{b}{a}h\right)$$

For this example, we can use the coordinates of the center $(-2, -3)$ and one of the corners of the rectangle $(0, 0)$ to write the equation of one of the asymptotes. The slope is

$$m = \frac{-3 - 0}{-2 - 0} = \frac{3}{2}$$

and the equation of one asymptote is

$$y - 0 = \frac{3}{2}(x - 0), \qquad \text{or} \qquad y = \frac{3}{2}x$$

Similarly, by using the center $(-2, -3)$ and an adjacent corner of the rectangle $(-4, 0)$, we can get the equation of the other asymptote. The slope is

$$m = \frac{-3 - 0}{-2 + 4} = \frac{-3}{2} = -\frac{3}{2}$$

and the equation of the other asymptote is

$$y - 0 = -\frac{3}{2}(x + 4) \qquad \text{or} \qquad y = -\frac{3}{2}x - 6$$

A sketch of this hyperbola with the asymptotes (dashed lines) is shown in Figure 7.8.

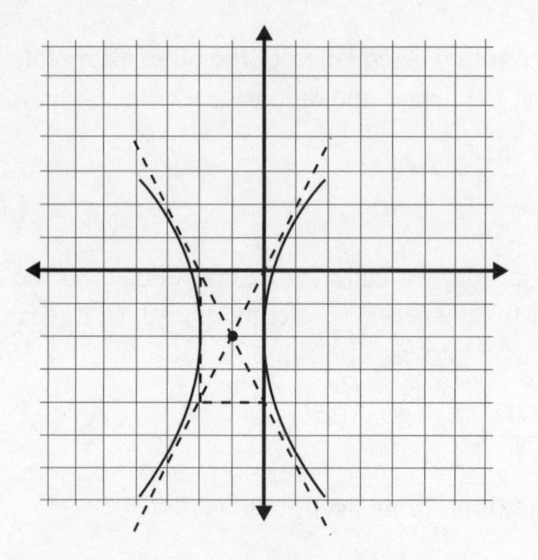

Figure 7.8

PARABOLA

If $A = 0$ or $C = 0$ (but not both) in the general form

$$Ax^2 + Bxy + Cy^2 + Ex + Fy + G = 0,$$

the conic is a parabola.

By definition, a parabola is the set of all points equidistant from a fixed point, called the focus, and a fixed line, called the directrix. The standard form is either $y = a(x - h)^2 + k$ (vertical parabola) or $x = a(y - k)^2 + h$ (horizontal parabola). See Figure 7.9. Sometimes these are also written as

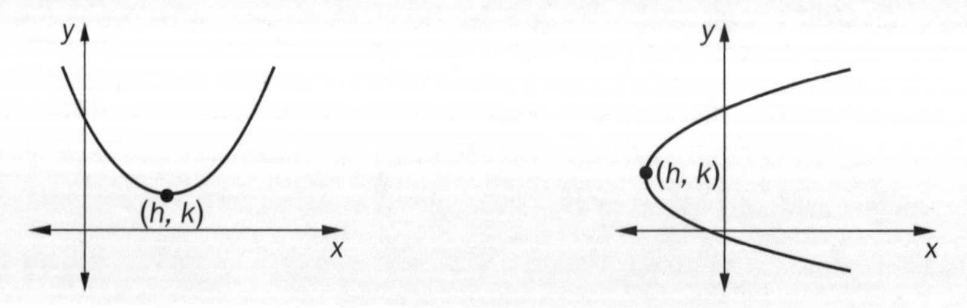

(a) Vertical parabola, $y = a(x - h)^2 + k$ (b) Horizontal parabola, $x = a(y - k)^2 + h$

Figure 7.9

$$y = \frac{1}{4p}(x-h)^2 + k \qquad \text{or} \qquad x = \frac{1}{4p}(y-k)^2 + h, \text{ where } 4p = \frac{1}{a}$$

The vertex of the parabola in both cases above is (h, k). For the parabola with the x-term squared, the axis of symmetry is $x = h$. The parabola opens up if $p > 0$ (or $a > 0$) and down if $p < 0$ (or $a < 0$).

The distance from the vertex to the focus, and hence from the vertex to the directrix, is p units along the axis of symmetry. This also gives us additional information that we can use to sketch the parabola. The line segment through the focus, parallel to the directrix and perpendicular to the axis of symmetry is called the focal chord, with length $|4p|$. Therefore, to sketch a parabola, we identify the vertex, the focus, and two additional points on the parabola that are also on the focal chord, or $2p$ units on either side of the focus.

For parabolas with the y-term squared, $x = \frac{1}{4p}(y - k)^2 + h$, the center is (h, k) and the axis of symmetry is $y = k$. The parabola opens to the right if $p > 0$ and to the left if $p < 0$. The distance from the vertex to the focus, and hence from the vertex to the directrix, is p units along the axis of symmetry. Figure 7.10 shows the geometry of a parabola opening to the right.

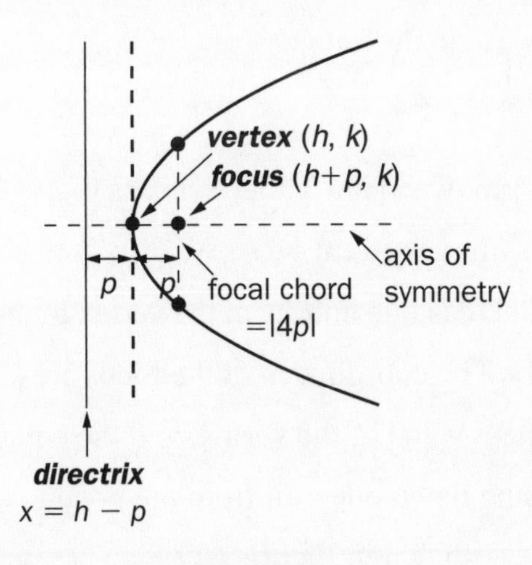

Figure 7.10

Example

Identify the vertex, the focus, and the equations of the directrix and axis of symmetry, and sketch the graph for the following equation:

$4x^2 - 16y - 12x = 23$

Solution

We first put the equation into one of the standard forms. Since only one of x and y is squared, we identify the conic as a parabola. We first rearrange the terms to place the x-terms on the right side of the equation, multiply both sides by (-1), and complete the square on the x-terms.

$$16y + 23 = 4x^2 - 12x$$

$$16y + 23 + 9 = 4\left(x^2 - 3x + \frac{9}{4}\right)$$

$$16(y + 2) = 4\left(x - \frac{3}{2}\right)^2$$

Dividing both sides by 16 and reducing the fractions, we get

$$y + 2 = \frac{1}{4}\left(x - \frac{3}{2}\right)^2 \text{ or}$$

$$y = \frac{1}{4}\left(x - \frac{3}{2}\right)^2 - 2$$

From this form, we identify the vertex as $\left(\frac{3}{2}, -2\right)$ and the equation of the axis of symmetry as $x = \frac{3}{2}$. We also notice that $4p = 4$, or $p = 1$. Therefore, there is one unit from the vertex to the focus, along the axis of symmetry. The coordinates of the focus are $\left(\frac{3}{2}, -1\right)$, going up one unit from the y-value of the vertex, and the equation of the directrix is $y = -3$, going down one unit from the y-value of the vertex. A sketch of the curve is shown in Figure 7.11.

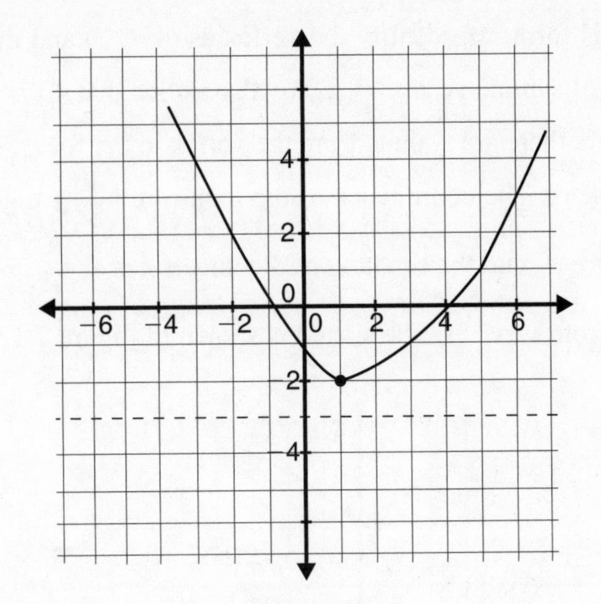

Figure 7.11

Example

Identify the vertex, the focus, and the equations of the directrix and axis of symmetry, and sketch the graph for the following equation:

$y^2 - 5x - 4y - 1 = 0$

Solution

We first put the equation into one of the standard forms. Since only one of x and y is squared, we identify the conic as a parabola. We complete the square on the y-terms to get

$$5x + 1 + 4 = (y^2 - 4y + 4)$$

$$5(x + 1) = (y - 2)^2$$

$$x + 1 = \frac{1}{5}(y - 2)^2$$

$$x = \frac{1}{5}(y - 2)^2 - 1$$

From this form, we identify the vertex as $(-1, 2)$ and the equation of the axis of symmetry as $y = 2$. We also notice that $4p = 5$, or $p = \frac{5}{4}$. Therefore, there are $\frac{5}{4}$ units from the vertex to the focus, along the axis of symmetry. The coordinates of the focus are $\left(\frac{1}{4}, 2\right)$, going $\frac{5}{4}$ units from $x = -1$, and the equation of the directrix is $x = -\frac{9}{4}$, going back $\frac{5}{4}$ units from $x = -1$. The sketch is shown in Figure 7.12.

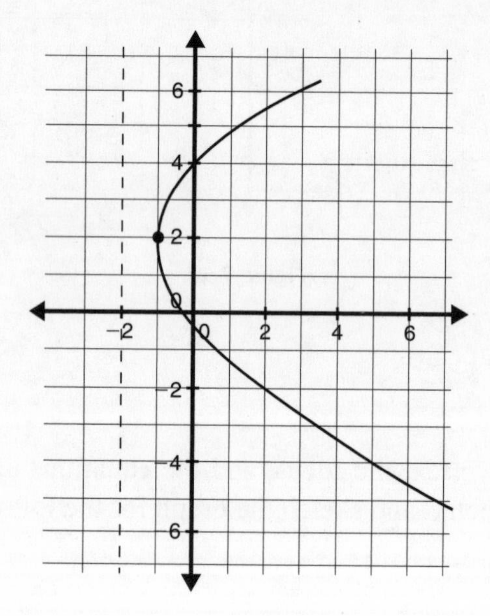

Figure 7.12

ROTATION OF AXES

If $B \neq 0$ in the general form $Ax^2 + Bxy + Cy^2 + Ex + Fy + G = 0$, then there is a rotation of axes that will transform the equation so that the xy term is eliminated. If the x- and y-axes are rotated through an angle θ, every point $P(x, y)$ in the xy-coordinate plane will have coordinates $P(x', y')$ in the $x'y'$-coordinate plane. These are related through the formulas

$$x = x'\cos\theta - y'\sin\theta \qquad \text{and} \qquad y = x'\sin\theta + y'\cos\theta$$

You can determine the angle θ through which the axes should be rotated by the formula $\cot(2\theta) = \frac{A - C}{B}$, given the general form $Ax^2 + Bxy + Cy^2 + Ex + Fy + G = 0$ with $B \neq 0$.

PRACTICE TEST 1

Notes and Reference Information

The following information will be available for reference during the exam.

1. Figures that accompany questions are intended to provide information useful in answering the questions. All figures lie in a plane unless otherwise indicated. The figures are drawn as accurately as possible EXCEPT when it is stated in a specific question that the figure is not drawn to scale. Straight lines and smooth curves may appear slightly jagged on the screen.

2. Unless otherwise specified, all angles are measured in radians, and all numbers used are real numbers. For some questions in this test, you may have to decide whether the calculator should be in radian mode or degree mode.

3. Unless otherwise specified, the domain of any function f is assumed to be the set of all real numbers x for which $f(x)$ is a real number. The range of f is assumed to be the set of all real numbers $f(x)$, where x is in the domain of f.

4. In this test, $\log x$ denotes the common logarithm of x (that is, the logarithm to the base 10) and $\ln x$ denotes the natural logarithm of x (that is, the logarithm to the base e).

5. The inverse of a trigonometric function f may be indicated using the inverse function notation f^{-1} or with the prefix "arc" (e.g., $\sin^{-1}x = \arcsin x$).

6. The range of $\sin^{-1}x$ is $\left[-\frac{\pi}{2}, \frac{\pi}{2}\right]$.

   The range of $\cos^{-1}x$ is $[0, \pi]$.

   The range of $\tan^{-1}x$ is $\left(-\frac{\pi}{2}, \frac{\pi}{2}\right)$.

7. Law of Sines: $\frac{a}{\sin A} = \frac{b}{\sin B} = \frac{c}{\sin C}$

   Law of Cosines: $c^2 = a^2 + b^2 - 2ab\cos C$

8. Sum and Difference Formulas:

   $\sin(\alpha + \beta) = \sin\alpha\cos\beta + \cos\alpha\sin\beta$

   $\sin(\alpha - \beta) = \sin\alpha\cos\beta - \cos\alpha\sin\beta$

   $\cos(\alpha + \beta) = \cos\alpha\cos\beta - \sin\alpha\sin\beta$

   $\cos(\alpha - \beta) = \cos\alpha\cos\beta + \sin\alpha\sin\beta$

CLEP Precalculus Practice Test 1

(Answer sheets appear in the back of this book.)

TOTAL TEST TIME: 90 Minutes
48 Questions

<u>**DIRECTIONS:**</u> This test consists of 48 questions: 25 questions in Section 1 and 23 questions in Section 2. All variables denote real numbers, except as noted otherwise. The domain of all functions consists of all numbers for which the function is defined, except as noted otherwise. Drawings may not be to scale. Some questions involve an analysis of graphs and require you to eliminate choices based on your analysis.

SECTION 1: 50 Minutes
25 Questions

<u>**DIRECTIONS:**</u> A graphing calculator will be available for Section 1 of this exam. Some questions will require you to select from five answer choices. Select the best answer. Other questions will require you to write a numerical answer in the box provided.

1. The June temperature of a central Texas town can be modeled by $T(t) = 30\sin\left(\frac{\pi}{12}t\right) + 68$, where t is the time in hours after noon. At what approximate time will the temperature first reach 95°?

(A) 3:30 A.M.

(B) 2:00 P.M.

(C) 3:00 P.M.

(D) 4:15 P.M.

(E) 8:30 A.M.

2. Solve $e^x + 3e^{-x} - 4 = 0$.

(A) 1 and -2.586

(B) 0 and 3.485

(C) 1 and -0.023

(D) 1 and 6.892

(E) 0 and 1.099

3. Which answer below is a possible solution for $\sin^2 3x - \sin 3x - 1 = 0$?

(A) 1.093 radians

(B) 5.014 radians

(C) 6.061 radians

(D) 2.920 radians

(E) 4.559 radians

4. Which of the following represents moving the graph of $y = x^2 - 8x + 15$ to the right three units and down two units?

(A) $y = x^2 + 5x + 13$

(B) $y = x^2 - 11x + 17$

(C) $y = x^2 + 3x - 2$

(D) $y = x^2 - 14x + 46$

(E) $y = -x^2 - 5x + 13$

5. The vertices of a hyperbola are $(0, -3)$ and $(-4, -3)$. The foci are $(-2 - \sqrt{13}, -3)$ and $(-2 + \sqrt{13}, -3)$. What is the x-coordinate of the center?

6. Given triangle ABC with $\angle A = 40°$, length of side $AB = 2$ units, and length of side $BC = 3$ units. What is an approximate measure of angle C?

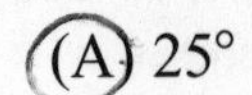

(A) 25°

(B) 48°

(C) 41°

(D) 37°

(E) This is an ambiguous case—two triangles are formed.

7.

x	-3	-2	1	4
$f(x)$	$-\frac{27}{2}$	-4	$\frac{1}{2}$	32

A power function can be defined as $f(x) = ax^p$, where a and p are constants. Using the values in the table above, find the value of a.

(A) $-\frac{1}{2}$

(B) $\frac{1}{2}$

(C) 2

(D) 3

(E) $\frac{2}{3}$

8. Given that f and g are both even functions, what must be true?

I. $f + g$ is even.

II. $f - g$ is odd.

III. $f \times g$ is neither.

IV. $f \times g$ cannot be determined.

(A) I only

(B) I and II only

(C) I, II, and III only

(D) I and IV only

(E) II and III only

9. Two birds take off from the same branch, with Bird A flying due west at 5 ft/sec and Bird B flying due south at 7 ft/sec. Let x and y be the distances flown by Bird A and Bird B, respectively. Assuming the birds continue in the given directions, find a function $f(t)$, where t is the time in seconds, that measures the distance between them.

(A) $12t$ (D) $74t$

(B) $2t$ (E) $t\sqrt{12}$

(C) $t\sqrt{74}$

10. Given $f(x) = \dfrac{3}{2-x} + 2x - 4$, what is the value of b so that $f(2-b) = -1$, if $b < 0$.

11. The decay of a radioactive substance can be modeled by $A(t) = A_0 e^{\frac{-t}{200}}$, where t is time in years, A_0 is the original amount, and A is the amount present at time t. How long before $\frac{1}{2}$ of the substance remains?

(A) 285.12 years (D) 234.23 years

(B) 138.63 years (E) 199.45 years

(C) 55.98 years

12. Find the equation of a circle in general form with center $(-1, -5)$ and tangent to the x-axis.

(A) $x^2 + y^2 + 2x + 10y + 1 = 0$

(B) $x^2 + y^2 - 2x - 10y + 1 = 0$

(C) $x^2 + y^2 + 2x + 10y + 25 = 0$

(D) $x^2 + y^2 - 2x - 10y - 25 = 0$

(E) $x^2 + y^2 - 2x - 10y = 1$

13. $$f(x) = \begin{cases} 3^x, & \text{if } x < 0 \\ 0, & \text{if } x = 0 \\ 2 - \frac{x}{3}, & \text{if } x > 0 \end{cases}$$

Given the function f defined above. What is the range of f?

(A) $\{y: y > 0\}$

(B) $\{y: -\infty < y < \infty\}$

(C) $\{y: 2 < y < \infty\}$

(D) $\{y: -\infty < y < 2\}$

(E) $\{y: y < 0\}$

14. Give the equations of the vertical asymptotes of the rational function

$$f(x) = \frac{x^2 + x - 2}{3x^2 - x - 10}$$

(A) $y = \frac{1}{3}$

(B) $y = 0$

(C) $x = 2,\ x = -\frac{5}{3}$

(D) $x = -2,\ x = \frac{5}{3}$

(E) $y = \frac{1}{5}$

15. Which of the following quadratic functions will have complex solutions?

I.

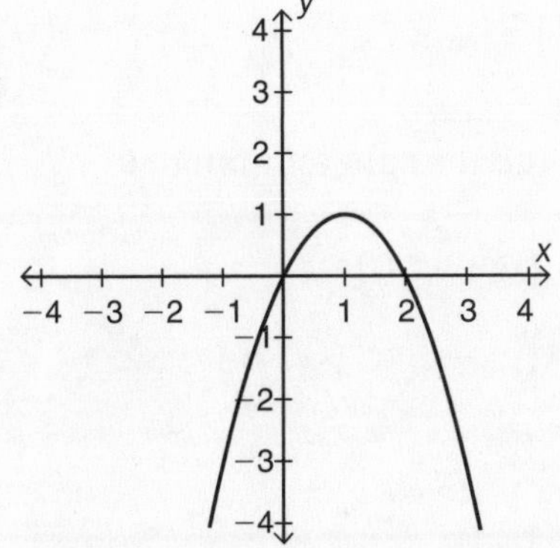

II.

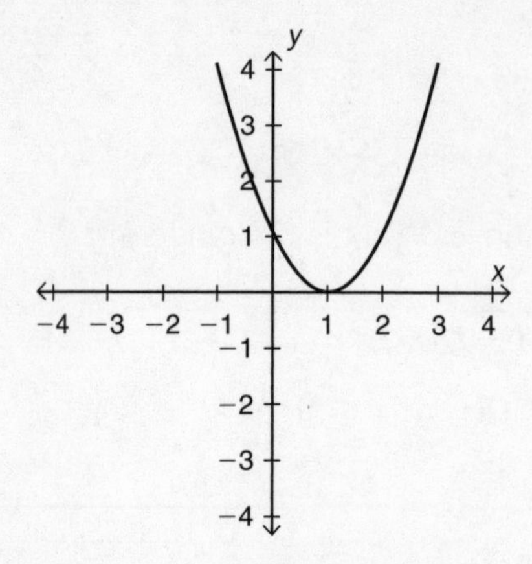

III.

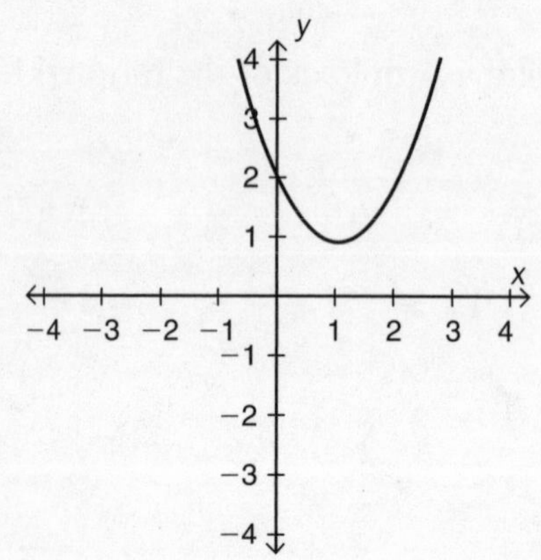

(A) I only
(B) I and III only
(C) III only
(D) II and III only
(E) I and II only

16. If $f(x) = ab^x$, $a < 0 < b$, what is the domain and range of f?

(A) $D = \{x: -\infty < x < \infty\}; R = \{y: y < 0\}$

(B) $D = \{x: -\infty < x < \infty\}; R = \{y: -\infty < y < \infty\}$

(C) $D = \{x: x > 0\}; R = \{y: y < 0\}$

(D) $D = \{x: x > 0\}; R = \{y: y > 0\}$

(E) $D = \{x: x \leq 0\}; R = \{y: y \leq 0\}$

17.

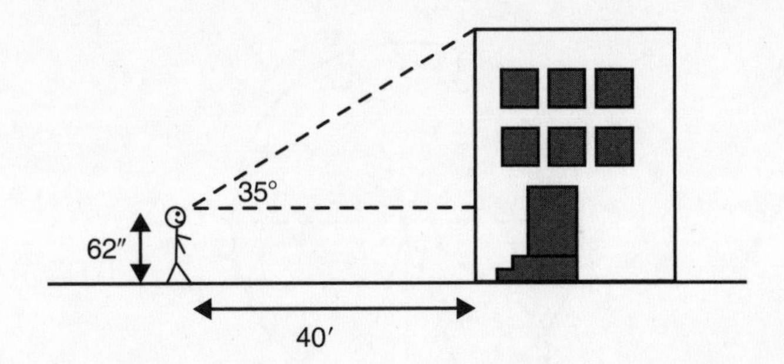

A student is asked to find the height of the school's gymnasium. He walks off 40 feet from the base of the building and measures an angle of 35° from his line of sight to the top of the building. If his height is 62 inches from his eyes to the ground, what is the height of the gym?

(A) 28 feet

(B) 33 feet, 2 inches

(C) 35 feet, 4 inches

(D) 90 feet

(E) 54 feet, 3 inches

18. A taxi charges a flat rate of \$2 for the first $\frac{1}{4}$ of a mile. Each quarter mile after the first is charged at \$.40, with the cost rounded up for any portion over $\frac{1}{4}$. You need to go to a store 1.6 miles away, but you have only \$7. Which choice below represents the change you would get if you paid for the trip plus a 15% tip?

(A) \$5.06

(B) \$1.62

(C) \$2.18

(D) \$1.94

(E) You do not have enough money for the trip and the tip.

19. Given $f(x) = \cos x$, $g(x) = x - 2$, $h(x) = e^{3x}$. Find the composition $f(g(h(x)))$.

(A) $e^{3(x-2)}$

(B) $e^{3\cos x - 2}$

(C) $\cos(e^{3x} - 2)$

(D) $\cos(e^{3x-2})$

(E) $e^{3\cos x} - 2$

20.

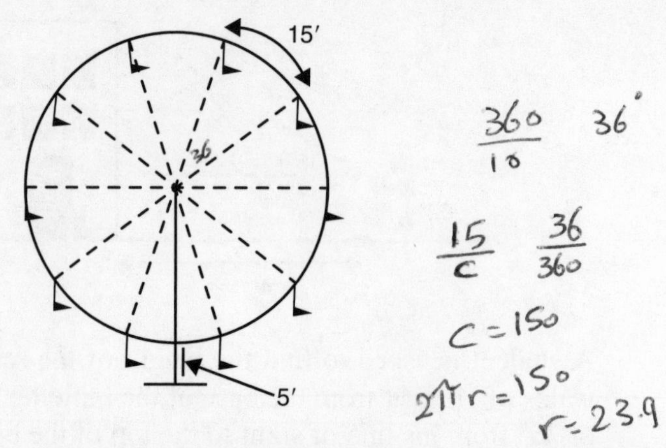

A children's Ferris wheel ride stands 5 feet above the ground and has 10 equally spaced baskets, each 15 feet apart along the arc. If a basket stops at the top of the Ferris wheel, approximately how far above the ground are the children in the basket?

(A) 65.15 feet
(B) 23.87 feet
(C) 52.74 feet
(D) 47.23 feet
(E) 35.86 feet

21. The quadratic equation

$$S(x) = -6.469x^2 + 1303x + b$$

models the 2003 median salary in the United States in terms of age, x. If the median salary of a 40-year-old was $31,450 in 2003, find b, and use the model to estimate the median salary of a 43-year-old in 2003.

(A) $40,298
(B) $35,184
(C) $38,296
(D) $33,748
(E) $42,682

22. It has been determined that the number of shoppers at a 24-hour megastore can be estimated by the equation $S(t) = -150\cos\left(\frac{5\pi}{6}t\right) + 350$, where S is the number of shoppers and t is the time in hours after midnight. How many shoppers should management expect at 8:30 A.M.?

(A) 345
(B) 275
(C) 205
(D) 405
(E) 495

23. Which represents the reflection of $y = f(x)$ about the x-axis?

(A) $y = f(x)$
(B) $y = -f(x)$
(C) $y = f(-x)$
(D) $y = |f(x)|$
(E) $y = f(|x|)$

24. Which of the following are vertices of the ellipse $9x^2 + y^2 - 54x + 6y + 81 = 0$?

(A) $(-3, 6), (-3, 0)$
(B) $(6, 0), (6, -3)$
(C) $(0, 3), (-6, 3)$
(D) $(3, 0), (3, -6)$
(E) $(0, -3), (6, 3)$

25. One model used to describe growth is $P(t) = P_0e^{kt}$, where P_0 is the initial population, t is time in hours, and k is a growth constant. If the population of a species is 2,000 at $t = 0$ and 7,500 after 10 hours, what is the approximate population after 16 hours using this model?

(A) 16,600
(B) 12,500
(C) 32,300
(D) 21,500
(E) 8,400

SECTION 2: 40 Minutes
23 Questions

DIRECTIONS: A graphing calculator will not be available for this section of the exam. Some questions will require you to select from five answer choices. Select the best answer. Other questions will require you to write a numerical answer in the box provided.

26. Find an equation of the circle in which the endpoints of a diameter are $P(2, 5)$ and $Q(4, 3)$.

(A) $(x - 3)^2 + (y - 4)^2 = 2$

(B) $(x - 4)^2 + (y - 3)^2 = 8$

(C) $(x - 2)^2 + (y - 2)^2 = 4$

(D) $(x - 1)^2 + (y - 7)^2 = 8$

(E) $(x - 2)^2 + (y - 5)^2 = 2$

27. Use the Remainder Theorem to find the remainder for:

$$\frac{4x^4 - 3x^2 + 5x - 2}{x+1}$$

28. What is the maximum value of $y = -\frac{1}{2}x^2 + 3x$?

(A) 0

(B) 3

(C) $\frac{9}{2}$

(D) $\frac{1}{2}$

(E) There is no maximum value.

29. If $\tan\theta = \frac{2}{3}$, for $\pi \leq \theta < \frac{3\pi}{2}$, what is $\cos\theta$?

(A) $\frac{3}{\sqrt{13}}$

(B) $\frac{-3}{\sqrt{13}}$

(C) $-\frac{\sqrt{13}}{3}$

(D) $\frac{-2\sqrt{13}}{3}$

(E) $\frac{3\sqrt{13}}{2}$

30. What is the period of $y = -4\sin\frac{x}{2}$?

(A) π

(B) $\frac{\pi}{4}$

(C) $\frac{\pi}{2}$

(D) 4π

(E) 2π

31. Which of the following is a solution to $\ln|x^2 - x - 6| - \ln|x + 2| = \ln\frac{1}{3}$?

(A) 3

(B) $\frac{3}{8}$

(C) $-\frac{8}{3}$

(D) 0

(E) $\frac{10}{3}$

32. Solve $\tan^2\theta = \sqrt{3}\tan\theta$ for θ in $[0, 2\pi)$.

(A) $0, \frac{\pi}{3}, \pi, \frac{4\pi}{3}$

(B) $0, \frac{2\pi}{3}, \pi, \frac{5\pi}{3}$

(C) $0, \frac{2\pi}{3}$

(D) $0, \frac{5\pi}{3}$

(E) $0, \frac{2\pi}{3}, \pi$

33.

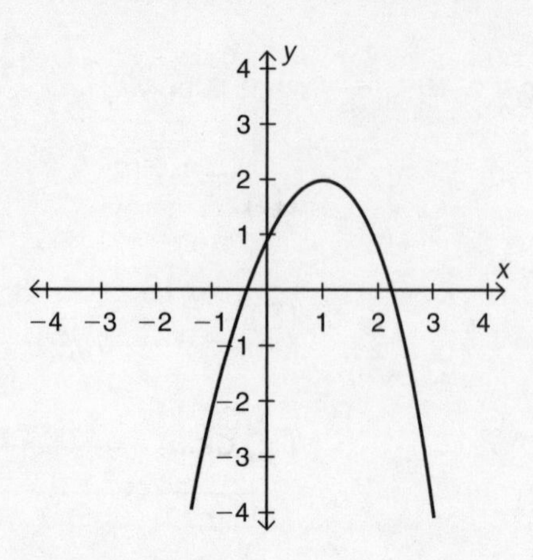

The graph above is a translation of the parent function $f(x) = x^2$ and is represented by which function?

(A) $f(x) = -x^2 + 2x$

(B) $f(x) = -x^2 + 2x + 1$

(C) $f(x) = -x^2 + 2x - 2$

(D) $f(x) = -x^2 - 2x + 1$

(E) $f(x) = -x^2 - 2x$

34. Find the zeros of the function $f(x) = 3^x \times 2x^2 - 3^x \times 7x - 3^{x+2}$.

(A) $0, \frac{9}{2}, 1$

(B) $\frac{9}{2}, -1$

(C) $0, -\frac{9}{2}, 1$

(D) $0, \frac{9}{2}, -1$

(E) $-\frac{9}{2}, 1$

35. What is the range of $y = 2^{3 - \cos x}$?

(A) $\{y: -1 \leq y \leq 1\}$

(B) $\{y: y \geq 0\}$

(C) $\{y: -\infty < y < \infty\}$

(D) $\{y: y > 0\}$

(E) $\{y: 4 \leq y \leq 16\}$

36. Given $y = 3x^2 - 2x + c$. If $c > 1$, how many x-intercepts are possible?

37. The graph of $x = |y| + 1$ lies in which quadrant(s)?

(A) quadrant I only

(B) quadrants I and II only

(C) quadrants I and III only

(D) quadrants I and IV only

(E) quadrants I, III, and IV only

38. Use the trigonometric addition formulas to evaluate cos 105°.

(A) $\dfrac{\sqrt{2}+\sqrt{6}}{4}$

(B) $\dfrac{\sqrt{6}-\sqrt{2}}{4}$

(C) $\dfrac{\sqrt{2}+\sqrt{6}}{2}$

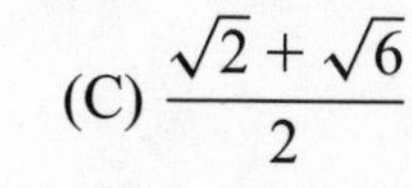

(D) $\dfrac{\sqrt{2}-\sqrt{6}}{4}$

(E) $\dfrac{\sqrt{2}-\sqrt{6}}{2}$

39.

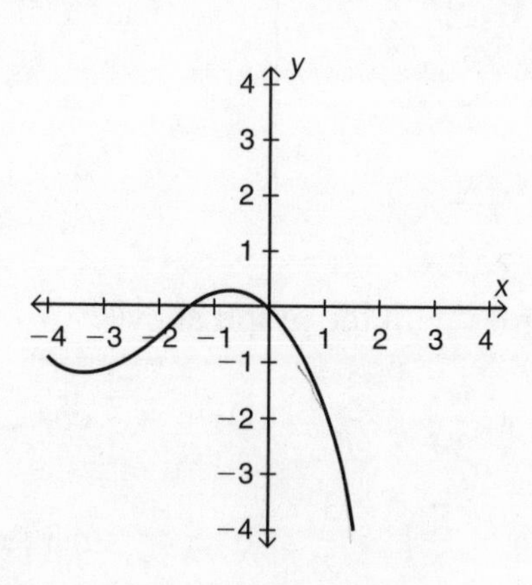

Identify the equation for the graph above.

(A) $y = e^x + \cos x$

(B) $y = \cos x - e^x$

(C) $y = \sin x + e^{-x}$

(D) $y = e^x - e^{-x}$

(E) $y = e^x + e^{-x}$

40. Solve $\log_x \left(\frac{1}{64}\right) = 2$ for x. $(x > 0)$

41. Given $f(x) = 2^x$ and $h(x) = x^2$. What will be true about $f(x)$ and $h(x)$ for all values of x in the interval $2 \leq x \leq 4$?

(A) $f(x) > h(x)$

(B) $f(x) < h(x)$

(C) $f(x) = h(x)$

(D) $f(x) \geq h(x)$

(E) $f(x) \leq h(x)$

42.

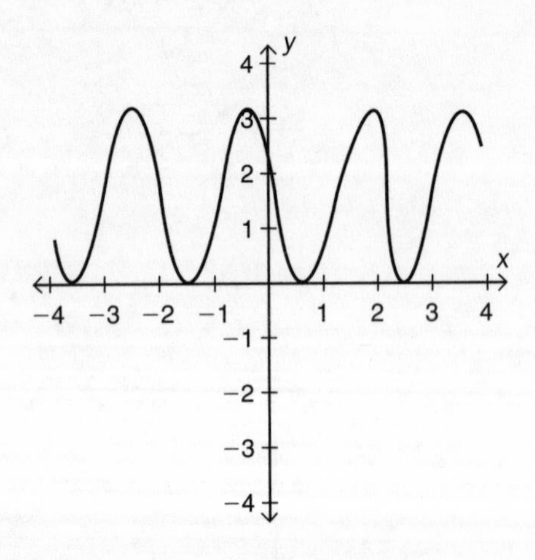

Which function goes with the graph above?

(A) $y = \frac{3}{2}\cos\left(x + \frac{\pi}{3}\right) + \frac{3}{2}$

(B) $y = \frac{3}{2}\sin 3\left(x - \frac{\pi}{3}\right) + \frac{3}{2}$

(C) $y = \frac{3}{2}\sin 3\left(x - \frac{\pi}{3}\right) - 1$

(D) $y = \frac{2}{3}\cos 2\left(x - \frac{\pi}{6}\right) - \frac{3}{2}$

(E) $y = \sin 3\left(x + \frac{\pi}{3}\right) + 1$

43. Solve $|3 - 2x| = |5 - 4x|$ for x.

$2x = 2$
$x = 1$

(A) 1

$-6x = 8$
$x = \frac{8}{-6}$

(B) $1, -\frac{4}{3}$

(C) $1, \frac{4}{3}$

(D) $-1, -\frac{4}{3}$

(E) $-1, \frac{4}{3}$

44. Which of the following is an equation of a parabola with a y-intercept of 2 and x-intercepts of 1 and 4?

(A) $y = x^2 + 4x - 2$

(B) $y = -\frac{1}{2}x^2 - \frac{3}{2}x + 2$

(C) $y = \frac{1}{2}x^2 + \frac{3}{2}x - 2$

(D) $y = \frac{1}{2}x^2 - \frac{5}{2}x + 2$

(E) $y = \frac{5}{2}x^2 - \frac{1}{2}x - 1$

$1x^2 - 5x + 4$
$(x-1)(x-4)$

$-x^2 - 3 + 2$
$x^2 + 3 - 2$
$(x\ \ )(x\ \ 2)$

45. Solve $8^{3-x} = 16^{5-x}$ for x.

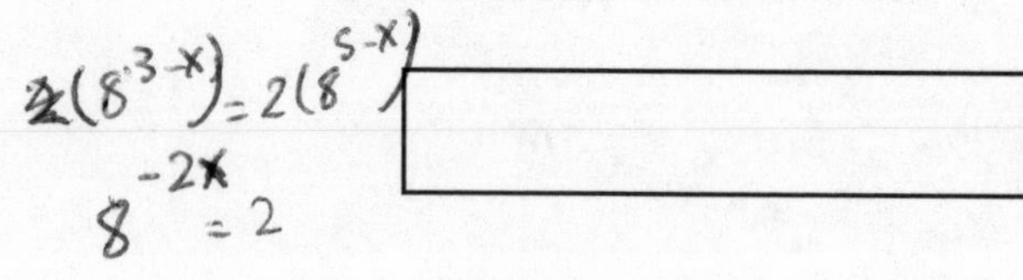

46. Which of the following is NOT equivalent to $\frac{1 - \tan^2\theta}{1 - \tan^4\theta}$?

(A) $\sin^2\theta$

(B) $\frac{1}{1 + \tan^2\theta}$

(C) $\frac{1}{\sec^2\theta}$

(D) $\cos^2\theta$

(E) $1 - \sin^2\theta$

$\frac{1 - \frac{\sin^2\theta}{\cos^2\theta}}{1 - \frac{\sin^4\theta}{\cos^4\theta}}$

47. Given $f(x) = -x^3$, consider the following functions:

I. $f(2x)$

II. $2f(x)$

III. $f\left(\frac{2}{x}\right)$

Which of the functions above are odd functions?

$F'(-x) = -F(x)$

(A) I only

(B) I and II only

(C) II only

(D) I and III only

(E) I, II, and III

48. Solve $6 \sin^{-1} \frac{x}{3} = \pi$ for x.

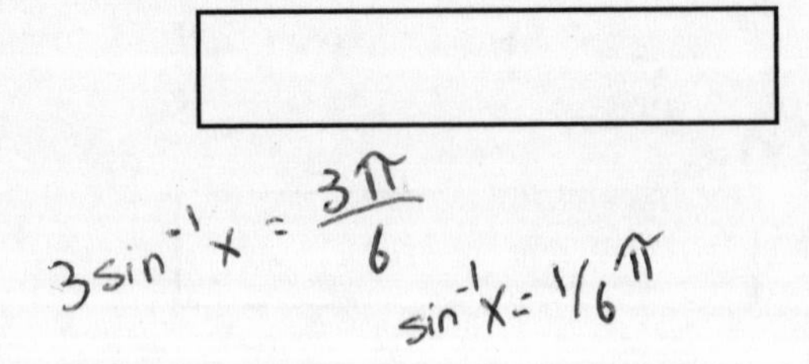

Answer Key
Practice Test 1

1. (D)
2. (E)
3. (C)
4. (D)
5. −2
6. (A)
7. (B)
8. (A)
9. (C)
10. −1
11. (B)
12. (A)
13. (D)
14. (C)
15. (C)
16. (A)
17. (B)
18. (D)
19. (C)
20. (C)
21. (D)
22. (E)
23. (B)
24. (D)
25. (A)
26. (A)
27. −6
28. (C)
29. (B)
30. (D)
31. (E)
32. (A)
33. (B)
34. (B)
35. (E)
36. 0
37. (D)
38. (D)
39. (B)
40. $\frac{1}{8}$
41. (E)
42. (B)
43. (C)
44. (D)
45. 11
46. (A)
47. (E)
48. $\frac{3}{2}$

Detailed Explanations of Answers Practice Test 1

1. **(D)**

$$T(t) = 30\sin\left(\frac{\pi}{12}t\right) + 68$$

We set up the equation and solve for t:

$$95 = 30\sin\left(\frac{\pi}{12}t\right) + 68$$

$$27 = 30\sin\left(\frac{\pi}{12}t\right)$$

$$\frac{27}{30} = \sin\left(\frac{\pi}{12}t\right)$$

$$.9 = \sin\left(\frac{\pi}{12}t\right)$$

$$t = \frac{(12)(\sin^{-1}.9)}{\pi} \approx 4.27 \text{ hours}$$

or about 4 hours and 15 minutes after noon (4:15 P.M.)

2. **(E)**

$$e^x + 3e^{-x} - 4 = 0$$

Multiply everything by e^x and rearrange the terms.

$$e^x(e^x + 3e^{-x} - 4 = 0)$$

$$e^{2x} + 3 - 4e^x = 0$$

$$e^{2x} - 4e^x + 3 = 0$$

This is in the form of a quadratic equation in e^x. We now factor and solve:

$$(e^x - 1)(e^x - 3) = 0$$

$$e^x - 1 = 0 \qquad e^x - 3 = 0$$

$$e^x = 1 \qquad e^x = 3$$

$$x = 0 \qquad x = 1.099$$

3. **(C)**

$\sin^2 3x - \sin 3x - 1 = 0$

Although this is a quadratic function in $\sin 3x$, it will not factor. We must use the quadratic formula to solve, with $a = 1, b = -1, c = -1$.

$$\sin 3x = \frac{1 \pm \sqrt{(-1)^2 - 4(1)(-1)}}{2(1)} = \frac{1 \pm \sqrt{5}}{2} \approx 1.618, \ -.6180$$

We discard the answer of 1.618 since it is out of range for the sine function. We now solve for x.

$$x \approx \frac{\sin^{-1}(-.6180)}{3} \approx -.2221$$

But, we were asked to find the answer for $0 \leq x < 2\pi$. The angle we found is in the fourth quadrant because it is a negative angle. We can find the equivalent positive angle by subtracting it from 2π:

$2\pi - .2221 \approx 6.061$ radians

4. **(D)**

$y = x^2 - 8x + 15$

We first put this equation into standard form by completing the square:

$y = (x^2 - 8x + \underline{\quad}) + 15 - \underline{\quad}$

$y = (x^2 - 8x + \underline{16}) + 15 - \underline{16}$

$y = (x - 4)^2 - 1$

To move three units to the right, and two units down, we will have

$y = (x - 4 - 3)^2 - 1 - 2$

$y = (x - 7)^2 - 3$

Multiplying this out, we have

$y = x^2 - 14x + 49 - 3$

$y = x^2 - 14x + 46$

5. **(−2)**

We do not need the information about the foci. The center is the midpoint between the vertices.

$$\left(\frac{0+(-4)}{2}, \frac{-3+(-3)}{2}\right) = (-2, -3)$$

6. **(A)**

We set this up using the Law of Sines:

$$\frac{\sin 40°}{3} = \frac{\sin C}{2}$$

Solving for $\sin C$ we get: $\sin C = \frac{2 \sin 40°}{3} \approx .4285$. Therefore $C \approx \sin^{-1}(.4285) \approx 25°$ *or* $C \approx \sin^{-1}(.4285) \approx 155°$. Both of these angles have a sine value of approximately .4285. But, the sum of angles A and C for the second choice (155°) is greater than 180°. Therefore, the only correct answer is $C \approx 25°$.

Note: Do not assume answer choice (E) right away.

7. **(B)**

To solve this problem, we can use a calculator to define the power function that generated the table values. Without a calculator, we can substitute the x and y values from the table into the general formula and solve for a and p. For this problem, the easiest point to use is $\left(1, \frac{1}{2}\right)$ in the function.

$$f(x) = ax^p$$

$$\frac{1}{2} = a(1)^p$$

$$\frac{1}{2} = a$$

8. **(A)**

Let $h(x) = f(x) + g(x)$. Since f and g are both even functions, then $f(-x) = f(x)$ and $g(-x) = g(x)$. Hence, $h(-x) = f(-x) + g(-x) = f(x) + g(x) = h(x)$. Therefore, their sum is even. This will also be true for their difference. Since their difference is even, then II is false. We now consider their product. Let $h(x) = f(x) \times g(x)$. Then $h(-x) = f(-x) \times g(-x) = f(x) \times g(x) = h(x)$. Therefore, the product is even also. This means that III and IV are false. The only one that is true is I.

9. **(C)**

Given that Bird A is flying at 5 ft/sec, then the bird's distance after t seconds is $5t$ feet. Similarly, Bird B has flown $7t$ feet after t seconds. The distance between them can be found with the Pythagorean Theorem:

$$d = \sqrt{(5t)^2 + (7t)^2}$$

$$d = \sqrt{25t^2 + 49t^2}$$

$$d = \sqrt{74t^2}, \text{ or } d = t\sqrt{74}$$

10. **(–1)**

$$f(x) = \frac{3}{2-x} + 2x - 4$$

$$f(2-b) = \frac{3}{2-(2-b)} + 2(2-b) - 4$$

$$= \frac{3}{b} + 4 - 2b - 4$$

$$= \frac{3-2b^2}{b}$$

(by finding a common denominator above and combining)

We are given that this is equal to -1, so the equation for b is

$$\frac{3-2b^2}{b} = -1$$

$$3 - 2b^2 = -b$$

Multiplying by -1 and setting this equal to 0, we get the quadratic equation

$$2b^2 - b - 3 = 0$$

We factor to solve.

$$(2b - 3)(b + 1) = 0$$

$$b = \frac{3}{2},\ b = -1$$

Since we are given that $b < 0$, the answer is $b = -1$.

11. **(B)**

$$A(t) = A_0 e^{\frac{-t}{200}}$$

For $\frac{1}{2}$ of the substance to remain, then the ratio $\frac{A}{A_0}$ must be $\frac{1}{2}$.

Solving for this ratio, we get $\frac{A}{A_0} = e^{\frac{-t}{200}}$ or $\frac{1}{2} = e^{\frac{-t}{200}}$.

We solve by taking the natural logarithm of both sides:

$$\ln\left(\frac{1}{2}\right) = \ln\left(e^{\frac{-t}{200}}\right)$$

$$\ln\left(\frac{1}{2}\right) = \frac{-t}{200}$$

$$t = -200 \ln\left(\frac{1}{2}\right) \approx 138.63 \text{ years}$$

12. **(A)**

Since the circle is tangent to the x-axis, the distance from the center to the x-axis is the radius, or 5 units. Using $(x - h)^2 + (y - k)^2 = r^2$, with $(h, k) = (-1, -5)$ and $r = 5$, we have

$$(x - (-1))^2 + (y - (-5))^2 = 5^2$$

$$(x + 1)^2 + (y + 5)^2 = 25$$

Multiplying this out, we have

$$x^2 + 2x + 1 + y^2 + 10y + 25 = 25$$

or

$$x^2 + y^2 + 2x + 10y + 1 = 0$$

13. **(D)**

$f(x) = 3^x$ for $x < 0, f(x) = 0$ for $x = 0$, and $f(x) = 2 - \frac{x}{3}$ for $x > 0$

For $x < 0$, the range will be $\{y: 0 < y < 1\}$.

For $x = 0$, the range will be $\{y: y = 0\}$.

For $x > 0$, the range will be $\{y: -\infty < y < 2\}$.

Therefore, taking the union of all these, we get the range $\{y: -\infty < y < 2\}$.

14. **(C)**

$$f(x) = \frac{x^2 + x - 2}{3x^2 - x - 10}$$

Vertical asymptotes occur where the denominator is equal to zero. We must also see whether there are "holes" in the graph, rather than vertical asymptotes, by checking for common variable factors in the numerator and denominator.

$$f(x) = \frac{(x-1)(x+2)}{(3x+5)(x-2)}$$

There are no holes in the graph, so we get the vertical asymptotes by setting each factor in the denominator equal to zero.

$3x + 5 = 0 \qquad x - 2 = 0$

$x = -\frac{5}{3} \qquad x = 2$

15. **(C)**

The graphs of functions with complex solutions do not cross or touch the x-axis. In other words, the graph is either completely above or below the x-axis. The graph in III is the only graph that does not touch or cross the x-axis.

16. **(A)**

$f(x) = ab^x, a < 0 < b$

Since there are no restrictions on x, the domain is all real values of x. Since b^x is always positive and $a < 0$ (given), the product ab^x is negative. Therefore, the range is all $y < 0$.

17. **(B)**

We set up a trigonometry equation: $\tan 35° = \frac{x}{40}$. Solving for x, we get

$$x = (40)(\tan 35°) \approx 28.$$

We now must add 62 inches to the 28 feet that we found. This will give us 33 feet, 2 inches.

18. **(D)**

Total cost of the trip: $\$2 + (\$.40)(6) = \$4.40$

15% tip: $= \$0.66$

Trip + tip: $= \$5.06$

Therefore, the change is $\$7.00 - \$5.06$, or $\$1.94$.

19. **(C)**

$$f(x) = \cos x,\ g(x) = x - 2,\ h(x) = e^{3x}$$

We first find $g(h(x))$.

$$g(h(x)) = g(e^{3x}) = e^{3x} - 2$$

Now, we find $f(g(h(x))) = f(e^{3x} - 2) = \cos(e^{3x} - 2)$

20. **(C)**

Since there are 10 equally spaced baskets, the central angle must be 36°, or $\frac{36\pi}{180}$ radians. The arc length is given by $s = r\theta$, where s is the arc length, r is the radius, and θ is the central angle.

$$r = \frac{s}{\theta} = \frac{15}{\frac{36\pi}{180}} \approx 23.87$$

The total distance will be the diameter plus the extra five feet above the ground. Therefore, the total distance is:

$$2 \times r + 5 \approx 2(23.87) + 5 = 52.74$$

Alternate solution:

Since there are 10 equally spaced baskets 15 feet apart, then the circumference is 150 feet.

$$C = \pi d$$
$$150 = \pi d$$
$$d = \frac{150}{\pi} \approx 47.74$$

To this diameter we add 5 feet to get $47.74 + 5 = 52.74$

21. **(D)**

$$S(x) = -6.469x^2 + 1303x + b$$

We let $x = 40$ and $f(x) = 31450$ and solve for b.

$$31450 = -6.469(40)^2 + 1303(40) + b$$
$$b \approx -10319.6$$

We now use this value of b in the equation and find $S(43)$.

$$S(x) = -6.469x^2 + 1303x - 10319.6$$
$$S(43) = -6.469(43)^2 + 1303(43) - 10319.6 \approx 33748$$

22. **(E)**

$$S(t) = -150\cos\left(\frac{5\pi}{6}t\right) + 350$$

We evaluate the function at $t = 8.5$.

$$S(8.5) = -150\cos\left(\frac{5\pi}{6}(8.5)\right) + 350 \approx 145 + 350 \approx 495$$

Note: Cosine is negative in that quadrant.

23. **(B)**

To reflect about the x-axis, we replace y with $(-y)$. Given $y = f(x)$ and replacing, we have $(-y) = f(x)$ or $y = -f(x)$.

24. **(D)**

$$9x^2 + y^2 - 54x + 6y + 81 = 0$$

We first put the equation into standard form:

$$9(x^2 - 6x + \underline{\quad}) + (y^2 + 6y + \underline{\quad}) = -81 + \underline{\quad} + \underline{\quad}$$

$$9(x^2 - 6x + \underline{\ 9\ }) + (y^2 + 6y + \underline{\ 9\ }) = -81 + \underline{\ 81\ } + \underline{\ 9\ }$$

$$9(x - 3)^2 + (y + 3)^2 = 9$$

$$\frac{(x-3)^2}{1} + \frac{(y+3)^2}{9} = 1$$

The major axis is the y-axis, so the vertices are ± 3 units away from the y-value of the center. We found the coordinates of the center as $(3, -3)$. Going ± 3 units from the y-coordinate of -3, we get $(3, 0)$ and $(3, -6)$.

25. **(A)**

$$P(t) = P_0 e^{kt}$$

In order to solve this problem we need to know the value of k. Since the population at $t = 0$ is 2000, $P_0 = 2000$, and our model becomes $P(t) = 2000e^{kt}$. We can now use the given information of $P(10) = 7500$ to substitute into the model:

$$7500 = 2000e^{10k}.$$

Solving for k, we get:

$$\frac{7500}{2000} = e^{10k}$$

Reducing and writing with logarithms, we have

$$\ln\left(\frac{75}{20}\right) = \ln\left(e^{10k}\right)$$

$$\ln\left(\frac{15}{4}\right) = 10k$$

$$k = \frac{\ln\left(\frac{15}{4}\right)}{10} \approx .132176$$

Our model is now $P(t) = 2000e^{.132176t}$. We evaluate at $t = 16$.

$$P(16) = 2000e^{.132176(16)} \approx 16576,$$

or approximately 16,600.

26. **(A)**

We use the distance formula to find the length of the diameter and the midpoint formula to get the coordinates of the center.

$$d = \sqrt{(4-2)^2 + (3-5)^2} = \sqrt{(2)^2 + (-2)^2} = \sqrt{8} = 2\sqrt{2}$$

Therefore, the length of the radius is $\frac{2\sqrt{2}}{2} = \sqrt{2}$. We now use the midpoint formula to get the coordinates of the center.

$$\left(\frac{4+2}{2}, \frac{3+5}{2}\right) \text{ or } (3,4)$$

Using the center as (3, 4) and the radius as $\sqrt{2}$, we have

$$(x-3)^2 + (y-4)^2 = (\sqrt{2})^2$$

or

$$(x-3)^2 + (y-4)^2 = 2$$

27. **(–6)**

$$\frac{4x^4 - 3x^2 + 5x - 2}{x+1}$$

The Remainder Theorem states: If a polynomial $P(x)$ is divided by the binomial $x - c$, the remainder is $P(c)$. For this problem $c = -1$ since we can write $x + 1$ as $x - (-1)$. We evaluate the polynomial $P(x) = 4x^4 - 3x^2 + 5x - 2$ at $x = -1$.

$$P(x) = 4x^4 - 3x^2 + 5x - 2$$

$$P(-1) = 4(-1)^4 - 3(-1)^2 + 5(-1) - 2 = -6$$

Therefore, the remainder is -6.

28. **(C)**

$$y = -\frac{1}{2}x^2 + 3x$$

Since the graph opens downward, we know there is a maximum value, which will be the y-value of the vertex. To find the vertex, we complete the square.

$$y = -\frac{1}{2}(x^2 - 6x + \underline{\quad}) + \underline{\quad}$$

$$y = -\frac{1}{2}(x^2 - 6x + \underline{9}) + \underline{\frac{9}{2}}$$

$$y = -\frac{1}{2}(x - 3)^2 + \frac{9}{2}$$

Since the coordinates of the vertex are $\left(3, \frac{9}{2}\right)$, the maximum value of the function is $\frac{9}{2}$.

29. **(B)**

$$\tan\theta = \frac{2}{3}, \text{ for } \pi \leq \theta < \frac{3\pi}{2}$$

Since $\tan\theta = \frac{y}{x}$ and we are given $\tan\theta = \frac{2}{3}$, then $\frac{y}{x} = \frac{2}{3}$. Since θ is in the third quadrant, x and y are negative. Therefore, $r = \sqrt{(-3)^2 + (-2)^2} = \sqrt{13}$ and $\cos = \frac{x}{r} = \frac{-3}{\sqrt{13}}$.

30. **(D)**

$$y = -4\sin\frac{x}{2}$$

For a sinusoidal function, the general form is $y = A\sin(B(\theta - h)) + k$, and the period is $\frac{2\pi}{|B|}$. Therefore, the period of the given function is $\frac{2\pi}{\frac{1}{2}} = 4\pi$.

31. **(E)**

$$\ln|x^2 - x - 6| - \ln|x + 2| = \ln\frac{1}{3}$$

We rewrite the equation using the rules of logarithms, specifically $\ln a - \ln b = \ln\frac{a}{b}$.

$$\ln\left|\frac{x^2 - x - 6}{x + 2}\right| = \ln\frac{1}{3}$$

We now rewrite this equation using base *e*.

$$e^{\ln\left|\frac{x^2 - x - 6}{x + 2}\right|} = e^{\ln\frac{1}{3}},$$

which simplifies to

$$\left|\frac{x^2 - x - 6}{x + 2}\right| = \frac{1}{3}$$

We factor the numerator and divide out the common factor of $(x + 2)$ in the numerator and denominator.

$$\left|\frac{(x + 2)(x - 3)}{x + 2}\right| = \frac{1}{3}$$

$$|x - 3| = \frac{1}{3}$$

$$x - 3 = \frac{1}{3} \quad \text{or} \quad x - 3 = -\frac{1}{3}$$

$$x = \frac{10}{3} \quad \text{or} \quad x = \frac{8}{3}$$

Both of these answers check in the original equation, but only $\frac{10}{3}$ is listed as an answer choice.

32. **(A)**

$$\tan^2\theta = \sqrt{3}\tan\theta$$

We set the equation equal to zero and factor to solve.

$$\tan^2\theta - \sqrt{3}\tan\theta = 0$$

$\tan\theta(\tan\theta - \sqrt{3}) = 0$

$\tan\theta = 0 \qquad \tan\theta - \sqrt{3} = 0$

$\theta = 0, \pi \qquad \tan\theta = \sqrt{3}$

$\theta = \frac{\pi}{3}, \frac{4\pi}{3}$

Therefore, the solutions are $\theta = 0, \frac{\pi}{3}, \pi, \frac{4\pi}{3}$.

33. **(B)**

From the parent function we move the vertex (h, k) from (0, 0) to (1, 2). Since the graph is opening downward, there is a negative coefficient of x^2. We can eliminate choice (A) and choice (E) since the x-intercepts for these functions do not match the graph. We can eliminate choice (C) since the y-intercept of that graph would be at -2. We are down to choice (B) or choice (D). We can put both equations into standard form to check the coordinates of the vertex.

Choice (D)	Choice (B)
$y = -x^2 - 2x + 1$	$y = -x^2 + 2x + 1$
$y = -(x + 1)^2 + 2$	$y = -(x - 1)^2 + 2$
Vertex would be at $(-1, 2)$.	Vertex would be at $(1, 2)$.

Choice (B) is the only choice that matches the given graph.

34. **(B)**

$f(x) = 3^x \times 2x^2 - 3^x \times 7x - 3^{x+2}$

To find the zeros of the function, we set it equal to zero and factor out the common factor of 3^x from all the terms.

$3^x \times 2x^2 - 3^x \times 7x - 3^{x+2} = 0$

$3^x(2x^2 - 7x - 3^2) = 0$

$3^x(2x - 9)(x + 1) = 0$

3^x will never equal zero. $\qquad 2x - 9 = 0 \qquad x + 1 = 0$

$x = \frac{9}{2} \qquad x = -1$

35. **(E)**

$$y = 2^{3 - \cos x}$$

We know that the range of cos x is $[-1, 1]$. If $\cos x = -1$, then $y = 2^{3 - \cos x} = 2^{3-(-1)} = 2^4 = 16$. If $\cos x = 1$, then $y = 2^{3 - \cos x} = 2^{3-(1)} = 2^2 = 4$.

Hence, the range of $y = 2^{3 - \cos x}$ is $4 \leq y \leq 16$.

36. **(0)**

$$y = 3x^2 - 2x + c$$

To determine the number of x-intercepts, we examine the discriminant of the quadratic equation $3x^2 - 2x + c = 0$. For this quadratic we have $a = 3$, $b = -2$, and $c = c$.

$$b^2 - 4ac = (-2)^2 - 4(3)(c) = 4 - 12c.$$

Since $c > 1$ (given), $4 - 12c < 0$, which means that the solutions to the quadratic equation are complex numbers. This indicates that there are no x-intercepts.

37. **(D)**

$$x = |y| + 1$$

This equation can be rewritten as $|y| = x - 1$. If $y > 0$, then the equation is $y = x - 1$. If $y < 0$, the equation is $(-y) = x - 1$ or $x + y = 1$. We can sketch these two lines, but we must remember that the function is defined only for $x \geq 1$. Therefore, only the parts of the lines for which $x \geq 1$ can be part of the graph of $x = |y| + 1$. This puts the graph in quadrants I and IV, where $x \geq 1$.

38. **(D)**

We use the trigonometric identity for the cosine of the sum of two angles. We separate 105° into the sum of 60° and 45°.

$$\cos(105°) = \cos(60° + 45°) = \cos 60° \cos 45° - \sin 60° \sin 45°$$

$$= \left(\frac{1}{2}\right)\left(\frac{\sqrt{2}}{2}\right) - \left(\frac{\sqrt{3}}{2}\right)\left(\frac{\sqrt{2}}{2}\right)$$

$$= \frac{\sqrt{2}}{4} - \frac{\sqrt{6}}{4} = \frac{\sqrt{2} - \sqrt{6}}{4}$$

39. **(B)**

Since (0, 0) is on the graph, we can eliminate options (A), (C), and (E). When $x = 0$, the function in option (A) is equal to 2; the function in option (C) is equal to 1, and the function in option (E) is equal to 2. This leaves us with either choice (B) or (D). We can eliminate choice (D) by examining the graph at $x = -1$. When we evaluate the function in (D) at $x = -1$, we get $e^{-1} - e^{1}$ or $\frac{1}{e} - e$, which is less than 0. But, the given graph is positive at $x = -1$. We have eliminated (A), (C), (D), and (E). We can check choice (B) at $x = -1$, $x = 0$, and $x = 1$. The value of the function in choice (B) is the only one that fits the graph, in terms of the function being positive, zero, and negative, respectively.

40. $\left(\frac{1}{8}\right)$

$$\log_x \left(\frac{1}{64}\right) = 2$$

We rewrite this logarithmic equation as an exponential equation:

$$x^2 = \frac{1}{64}$$

We now just take the square root, which must be positive, since $x > 0$ is given.

41. **(E)**

$$f(x) = 2^x \text{ and } h(x) = x^2$$

At $x = 2$ and $x = 4$, the functions are equal. To check whether there are no other points of intersection between the 2 and 4, we set the functions equal to each other.

$$2^x = x^2$$

Taking the logarithm of both sides, we get

$$\ln 2^x = \ln x^2$$

or

$$x \times \ln 2 = 2 \times \ln x$$

We can write this as:

$$\frac{x}{\ln x} = \frac{2}{\ln 2}$$

This is true only for $x = 2$ and $x = 4$. Since there are no points of intersection between 2 and 4, we know the graphs will not intersect and will not change from being "greater than" to being "less than" in that interval. We use 3 as a test point in the interval to see which function is greater at $x = 3$.

$$f(3) = 2^3 = 8$$

$$h(3) = 3^2 = 9$$

Therefore, in the interval $2 \leq x \leq 4, f(x) \leq h(x)$.

42. **(B)**

We recognize this as a sinusoidal function of the form $y = A \sin(B(\theta - h)) + k$ or $y = A \cos(B(\theta - h)) + k$. From the graph, we can see that the amplitude is $\frac{3}{2}$ and that the horizontal line through the middle of the graph is at $y = \frac{3}{2}$. Therefore, $A = \frac{3}{2}$ and $k = \frac{3}{2}$, which eliminates choices (C), (D), and (E). We then notice that there are three cycles in a 2π interval, which indicates that $B = 3$. This eliminates choice (A). Therefore, only choice (B) fits the graph.

43. **(C)**

$$|3 - 2x| = |5 - 4x|$$

We must consider two cases.

<u>Case 1</u>	<u>Case 2</u>
$3 - 2x = 5 - 4x$	$3 - 2x = -(5 - 4x)$
$2x = 2$	$3 - 2x = -5 + 4x$
$x = 1$	$8 = 6x$ or $x = \frac{4}{3}$

44. **(D)**

Since the y-intercept is 2 and the x-intercepts are 1 and 4, then we know (0, 2), (1, 0), and (4, 0) are points on the parabola (by definition of intercepts). We now put these values of x and y into the general form of a parabola: $y = ax^2 + bx + c$.

For (0, 2): $2 = a(0)^2 + b(0) + c$

$c = 2$

For (1, 0): $0 = a(1)^2 + b(1) + 2$

$a + b = -2$

For (4, 0): $0 = a(4)^2 + b(4) + 2$

$16a + 4b = -2$

We now solve these last two equations in two unknowns by multiplying $(a + b = -2)$ by -4 and combining to get:

$$-4a - 4b = 8$$

$$16a + 4b = -2$$

Combining, we get: $12a = 6$ or $a = \frac{1}{2}$. Since $a + b = -2$, $b = -\frac{5}{2}$.

Therefore, the answer is the equation of the parabola with $a = \frac{1}{2}$, $b = -\frac{5}{2}$, and $c = 2$, which is

$$y = \frac{1}{2}x^2 - \frac{5}{2}x + 2$$

45. **(11)**

$$8^{3-x} = 16^{5-x}$$

To solve, we express both sides of the equation in base 2.

$$(2^3)^{3-x} = (2^4)^{5-x}$$

$$2^{9-3x} = 2^{20-4x}$$

Since the bases are equal, the exponents will be equal. We set the exponents equal to each other and solve for x.

$$9 - 3x = 20 - 4x$$

$$x = 11$$

46. **(A)**

We simplify $\frac{1 - \tan^2 \theta}{1 - \tan^4 \theta}$ by factoring the denominator and dividing common factors.

$$\frac{1 - \tan^2 \theta}{1 - \tan^4 \theta} = \frac{1 - \tan^2 \theta}{(1 - \tan^2 \theta)(1 + \tan^2 \theta)}$$

$$= \frac{1}{1 + \tan^2 \theta} \text{ (choice B)}$$

$$= \frac{1}{\sec^2 \theta} \quad \text{(choice C)}$$

$$= \cos^2 \theta \quad \text{(choice D)}$$

$$= 1 - \sin^2 \theta \text{ (choice E)}$$

The only choice that is not equivalent is (A).

47. **(E)**

$$f(x) = -x^3$$

Recall that a function is odd if $f(-x) = -f(x)$. We now check each of the functions that are given.

We check $f(2x)$ to see whether $f(-(2x)) = -f(2x)$.

$$f(2x) = -(2x)^3 = -8x^3$$

$$-f(2x) = -(-(2x)^3) = 8x^3$$

$$f(-2x) = -(-2x)^3 = 8x^3 = -f(2x)$$

which verifies it is odd.

We check $2f(x)$ to see if $2f(-x) = -2f(x)$

$$2f(x) = 2(-x^3) = -2x^3$$

$$-2f(x) = -2(-x^3) = 2x^3$$

$$2f(-x) = 2(-(-x^3)) = 2x^3 = -2f(x)$$

which verifies it is odd.

We check $f\left(\frac{2}{x}\right)$ to see if $f\left(-\frac{2}{x}\right) = -f\left(\frac{2}{x}\right)$.

$$f\left(\frac{2}{x}\right) = -\left(\frac{2}{x}\right)^3 = -\frac{8}{x^3}$$

$$-f\left(\frac{2}{x}\right) = -\left[-\left(\frac{2}{x}\right)^3\right] = \frac{8}{x^3}$$

$$f\left(-\frac{2}{x}\right) = -\left(-\frac{2}{x}\right)^3 = \frac{8}{x^3} = -f\left(\frac{2}{x}\right)$$

which verifies it is odd.

All are odd functions, which is choice (E).

48. $\left(\mathbf{\frac{3}{2}}\right)$

$$6\sin^{-1}\frac{x}{3} = \pi$$

We rewrite the equation.

$$\sin^{-1}\frac{x}{3} = \frac{\pi}{6}$$

This is equivalent to:

$$\sin\frac{\pi}{6} = \frac{x}{3}$$

since $\sin^{-1} a = b \quad \Rightarrow \quad \sin b = a$

But $\sin\frac{\pi}{6} = \frac{1}{2}$, so we have $\frac{x}{3} = \frac{1}{2}$ or $x = \frac{3}{2}$

PRACTICE TEST 2

Notes and Reference Information

The following information will be available for reference during the exam.

1. Figures that accompany questions are intended to provide information useful in answering the questions. All figures lie in a plane unless otherwise indicated. The figures are drawn as accurately as possible EXCEPT when it is stated in a specific question that the figure is not drawn to scale. Straight lines and smooth curves may appear slightly jagged on the screen.
2. Unless otherwise specified, all angles are measured in radians, and all numbers used are real numbers. For some questions in this test, you may have to decide whether the calculator should be in radian mode or degree mode.
3. Unless otherwise specified, the domain of any function f is assumed to be the set of all real numbers x for which $f(x)$ is a real number. The range of f is assumed to be the set of all real numbers $f(x)$, where x is in the domain of f.
4. In this test, $\log x$ denotes the common logarithm of x (that is, the logarithm to the base 10) and $\ln x$ denotes the natural logarithm of x (that is, the logarithm to the base e).
5. The inverse of a trigonometric function f may be indicated using the inverse function notation f^{-1} or with the prefix "arc" (e.g., $\sin^{-1}x = \arcsin x$).
6. The range of $\sin^{-1}x$ is $\left[-\frac{\pi}{2}, \frac{\pi}{2}\right]$.

   The range of $\cos^{-1}x$ is $[0, \pi]$.

   The range of $\tan^{-1}x$ is $\left(-\frac{\pi}{2}, \frac{\pi}{2}\right)$.
7. Law of Sines: $\frac{a}{\sin A} = \frac{b}{\sin B} = \frac{c}{\sin C}$.

   Law of Cosines: $c^2 = a^2 + b^2 - 2ab\cos C$.
8. Sum and Difference Formulas:

   $\sin(\alpha + \beta) = \sin\alpha\cos\beta + \cos\alpha\sin\beta$

   $\sin(\alpha - \beta) = \sin\alpha\cos\beta - \cos\alpha\sin\beta$

   $\cos(\alpha + \beta) = \cos\alpha\cos\beta - \sin\alpha\sin\beta$

   $\cos(\alpha - \beta) = \cos\alpha\cos\beta + \sin\alpha\sin\beta$

CLEP Precalculus
Practice Test 2

(Answer sheets appear in the back of this book.)

TOTAL TEST TIME: 90 Minutes
48 Questions

<u>DIRECTIONS:</u> This test consists of 48 questions: 25 questions in Section 1 and 23 questions in Section 2. All variables denote real numbers, except as noted otherwise. The domain of all functions consists of all numbers for which the function is defined, except as noted otherwise. Drawings may not be to scale. Some questions involve an analysis of graphs and require you to eliminate choices based on your analysis.

SECTION 1: 50 Minutes
25 Questions

<u>DIRECTIONS:</u> A graphing calculator will be available for Section 1 of this exam. Some questions will require you to select from five answer choices. Select the best answer from the five choices. Other questions will require you to write a numerical answer in the box provided.

1. Solve $\frac{1}{2}\ln(3x+4) - 1 = \ln 2$ for x.

(A) 3.829

(B) 0

(C) 1.342

(D) 8.519

(E) 11.986

2. A square with side c is in the first quadrant, bounded by the x-axis and the y-axis, with a vertex at the origin. A line $y = -x + b$, $b > 0$, contains the point (c, c). Find an expression that represents the area bounded by the axes and the line enclosed in the first quadrant minus the area of the square.

   (A) $\frac{b^2}{2} - c^2$

   (B) $\frac{b^2 - c^2}{2}$

   (C) $\frac{c^2 - b^2}{2}$

   (D) $b^2 - c^2$

   (E) $c^2 - b^2$

3. Which of the following is a solution for $3\sin^2 4x + 3\sin 4x = 1$?

   (A) 3.502 radians

   (B) 2.841 radians

   (C) −2.136 radians

   (D) 1.053 radians

   (E) .067 radians

4. $x^2 + y^2 - 2x - 10y = 10$ and $x^2 + y^2 - 8x - 2y + 1 = 0$

   Find the distance between the centers of the two circles whose equations are given above.

5. What variation of the function $y = x^2$ will produce a graph that is shifted to the right seven units, down one unit, is four times wider than the curve produced by the parent equation, and opens downward?

   (A) $y = -\frac{1}{4}(x - 7)^2 - 1$

   (B) $y = -4(x - 7)^2 + 1$

   (C) $y = -4(x - 1)^2 + 7$

   (D) $y = -\frac{1}{4}(x + 1)^2 - 7$

   (E) $y = -(4x + 7)^2 - 1$

6. Which functions listed below are odd functions?

I. $f(x) = x^2 \sin x \cos x$

II. $f(x) = \dfrac{2}{\sin x}$

III. $f(x) = \ln|2 - x|$

IV. $f(x) = \dfrac{e^x}{x^2}$

(A) I only

(B) I and II only

(C) III and IV only

(D) I, II, and IV only

(E) IV only

7.

Town	Population	Annual Growth Rate
A	35,000	4%
B	26,000	7.5%

The table above gives the population and annual growth rate for two towns. If the growth continues at this rate, approximately when will the population of Town B catch up to that of Town A?

(A) 5 years

(B) 27 years

(C) 15 years

(D) 9 years

(E) 2 years

8.

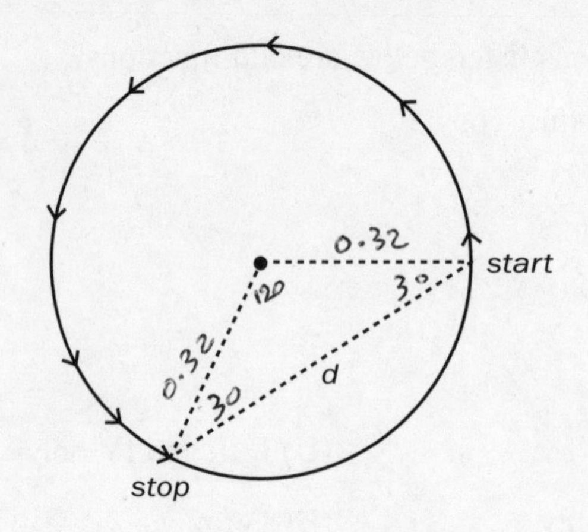

Suppose you walk $\frac{2}{3}$ around a 2-mile circular track. What is the straight-line distance (d) you have walked, relative to your starting point?

(A) 1.33 miles

(B) .86 miles

(C) .32 miles

(D) 1.02 miles

(E) .55 miles

9.

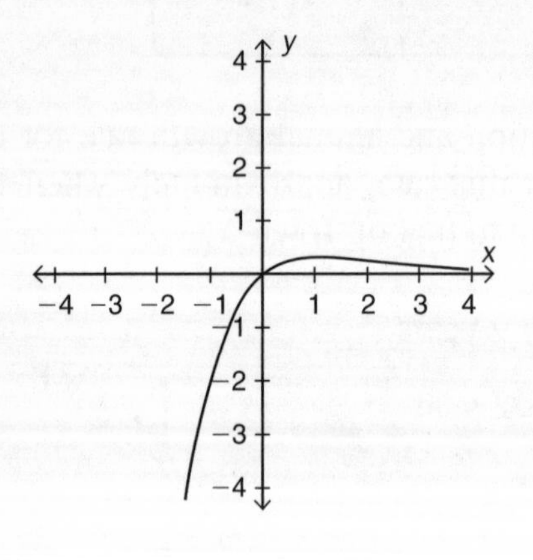

Function f

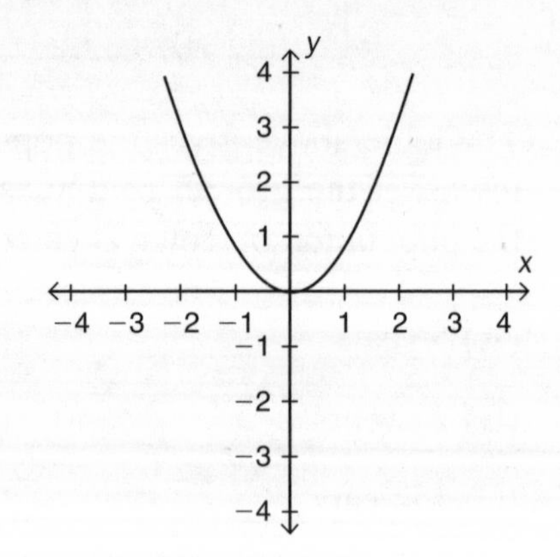

Function g

Let f and g be the functions shown above. Which function h could represent their product?

(A) $h(x) = \dfrac{1}{x}$

(B) $h(x) = e^x$

(C) $h(x) = xe^2$

(D) $h(x) = \dfrac{x^3}{e^x}$

(E) $h(x) = \dfrac{e^x}{x}$

10. A hyperbola's vertices are located at $(0, -3)$ and $(0, 3)$ and its asymptote equations are $y = \pm\dfrac{3}{4}x$. What is an equation of the hyperbola?

(A) $9y^2 - 16x^2 = 144$

(B) $16y^2 - 9x^2 = 144$

(C) $16y^2 - 9x^2 = 1$

(D) $9y^2 - 16x^2 = 1$

(E) $9x^2 - 16y^2 = 144$

11.

x	$f(x)$	$g(x)$	$f \circ g$
0	0	3	5
1	2	2	?
2	4	1	2
3	5	0	0

Given the table of values above for f, g, and $f \circ g$, what is the missing element in the table?

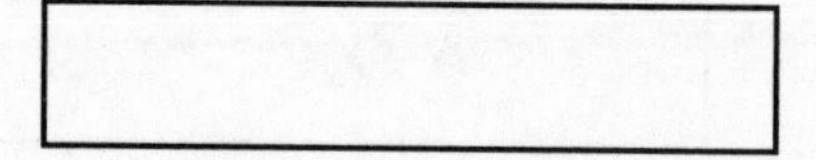

12.

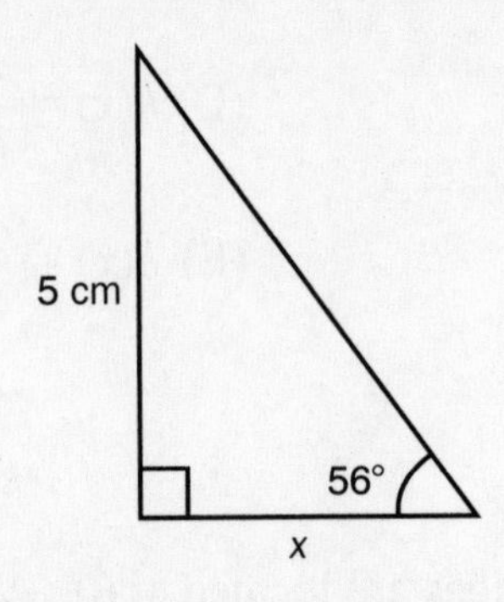

Find an approximate value for x in the above figure.

(A) 4.325 cm

(B) 3.373 cm

(C) 5.482 cm

(D) 4.089 cm

(E) 2.812 cm

13. Which of the following are coordinates of one of the foci of the ellipse given by $25x^2 + 4y^2 + 100x + 16y + 16 = 0$?

(A) $(-2, 3)$

(B) $(-2, -2 + \sqrt{21})$

(C) $(-2, -7)$

(D) $(-4, -2)$

(E) $(2, -2 + \sqrt{21})$

14. Given $f(x) = ab^x$, $b > 0$. If $f(1) = -6$ and $f(-1) = -\frac{3}{2}$, find a.

15. The population of deer in a 100-acre enclosure varies according to $P(t) = 750 \sin \frac{\pi}{6} t + 800$, where t is time in months. What approximate date would you expect the deer population to reach 1500 if there were 800 deer on January 1, 2008?

(A) March 10, 2008

(B) January 1, 2009

(C) May 30, 2009

(D) July 20, 2008

(E) September 20, 2008

16. What must be true about m and b if $y = x^2$ and $y = mx + b$ do not intersect?

(A) $m > 0$ and $b > 0$

(B) $2b > m^2$

(C) $m < 0$ and $b > 0$

(D) $m > 1$ and $b > 0$

(E) $b < 0$ and $m^2 + 4b < 0$

17. You wish to get a car loan for \$20,000 for 60 months. What is the difference in monthly payments between a loan of 8.9% APR and 5.6% APR? The monthly payment can be found by using the formula

$$p = \frac{\frac{lr}{12}}{1 - (1 + \frac{r}{12})^{-m}},$$

where p is the monthly payment, l is the loan amount, r is the annual interest rate (APR), and m is the number of months of the loan.

(A) \$123.08

(B) \$31.25

(C) \$62.90

(D) \$109.18

(E) \$74.32

18. A certain bacteria doubles every 8 hours and grows according to the model $N(t) = N_0e^{kt}$. If N_0 is the original amount, estimate the number of bacteria after 30 hours.

(A) $N_0 \times (7.07)$

(B) $N_0 \times (2.56)$

(C) $N_0 \times (32.72)$

(D) $N_0 \times (13.45)$

(E) $N_0 \times (1.09)$

19. Abby and Cammy are walking through an airport at the same rate of 3 ft/sec, with Abby 15 feet ahead of Cammy. Cammy then steps onto a 100-foot-long moving walkway, which is moving at 2.5 ft/sec and continues walking at her same rate. At what rate should Abby begin walking to reach the end of the walkway at the same time as Cammy?

(A) 3.2 ft/sec

(B) 2.5 ft/sec

(C) 5.2 ft/sec

(D) 4.7 ft/sec

(E) 1.9 ft/sec

20. $f(x) = \dfrac{x^2 + x - 2}{3x^2 - x - 10}$

What is an equation of a horizontal asymptote for the rational function given above?

(A) $y = \dfrac{1}{3}$

(B) $y = 0$

(C) $x = 2,\ x = -\dfrac{5}{3}$

(D) $x = -2,\ x = \dfrac{5}{3}$

(E) $y = \dfrac{1}{5}$

21. Given the functions $f(x) = e^x - x^2$ and $g(x) = \sqrt{x-2}$, which expression below represents $g(f(2))$?

(A) 0

(B) 1

(C) $\sqrt{e^2 - 6}$

(D) -1

(E) $e^{\sqrt{x-2}} - x + 2$

22. What must be true about b and c so that the graph of $y = x^2 + bx + c$ passes through only two quadrants?

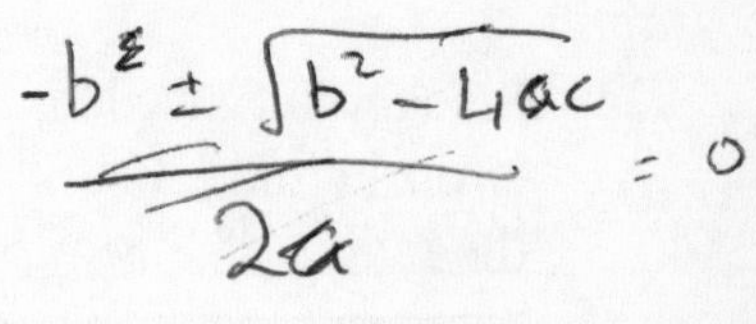

(A) $c \leq \frac{b^2}{4}$

(B) $c \geq \frac{b^2}{4}$

(C) $c = \frac{b^2}{2}$

(D) $c > \frac{b^2}{2}$

(E) Not enough information is given.

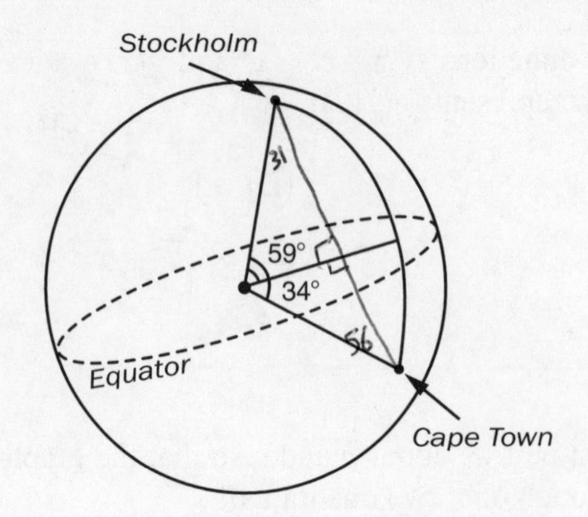

23. Stockholm, Sweden, is on latitude 59° N and Cape Town, South Africa, is on latitude 34° S. Both are on the same longitude. Assuming that the radius of Earth is approximately 3900 miles, estimate the distance between Stockholm and Cape Town.

(A) 4850 miles
(B) 3395 miles
(C) 6330 miles
(D) 10,455 miles
(E) 8840 miles

24. Find an inverse function for $f(x) = \dfrac{x}{2x-1}$.

(A) $f^{-1}(x) = -\dfrac{x}{2x-1}$
(B) $f^{-1}(x) = \dfrac{x}{2x+1}$
(C) $f^{-1}(x) = \dfrac{1}{2x-1}$
(D) $f^{-1}(x) = \dfrac{x}{2x-1}$
(E) $f^{-1}(x) = \dfrac{2x-1}{x}$

25. The number of bacteria can be modeled by $P = P_0 e^{.035t}$, where P_0 is the initial amount at $t = 0$ and t is measured in hours. How long will it take 30,000 bacteria to grow to 800,000?

(A) 231.5 hours
(B) 26.7 hours
(C) 115.9 hours
(D) 93.8 hours
(E) 155.2 hours

SECTION 2: 40 Minutes
23 Questions

DIRECTIONS: A graphing calculator will not be available for this section of the exam. Some questions will require you to select from five answer choices. Select the best answer. Other questions will require you to write a numerical answer in the box provided.

26. Solve $\log_{\sqrt{5}} 125 = 2x$ for x.

$\sqrt{5}^{2x} = 125$

$= 3$

27. Which of the following is NOT equivalent to $\dfrac{\sin\theta}{1+\cos\theta} + \dfrac{1+\cos\theta}{\sin\theta}$?

(A) $2\csc\theta$

(B) $\dfrac{2}{\sin\theta}$

(C) $\dfrac{2\sin\theta}{\tan\theta}$

(D) $\dfrac{2\tan\theta}{\sin^2\theta\sec\theta}$

(E) $\dfrac{2\sec\theta}{\tan\theta}$

28. What is the minimum value of $y = \dfrac{1}{2}\sin 3x + \dfrac{3}{2}$?

(A) 1

(B) $\dfrac{3}{2}$

(C) $-\dfrac{1}{2}$

(D) -1

(E) 0

29. What is the domain of $y = 2\sin^{-1}\left(\frac{x}{2}\right)$?

(A) $\{x: -1 \le x \le 1\}$

(B) $\{x: -2 \le x \le 2\}$

(C) $\left\{x: -\frac{1}{2} \le x \le \frac{1}{2}\right\}$

(D) $\{x: 0 \le x \le 1\}$

(E) $\left\{x: -\frac{1}{4} \le x \le \frac{1}{4}\right\}$

30. Which of the following is a factor of $x^3 + 4x^2 - 23x + 6$?

(A) $(x - 1)$

(B) $(x - 3)$

(C) $(x + 1)$

(D) $(x - 2)$

(E) $(x + 2)$

31. Given $\cos\theta = \frac{2}{3}$ for $\frac{3\pi}{2} \le \theta < 2\pi$. Find $\sin\theta$ and $\tan\theta$.

(A) $\sin\theta = \frac{\sqrt{5}}{3}, \tan\theta = \frac{\sqrt{5}}{2}$

(B) $\sin\theta = \frac{\sqrt{5}}{3}, \tan\theta = -\frac{\sqrt{5}}{2}$

(C) $\sin\theta = \frac{\sqrt{5}}{2}, \tan\theta = -\sqrt{5}$

(D) $\sin\theta = -\frac{\sqrt{5}}{3}, \tan\theta = -\frac{\sqrt{5}}{2}$

(E) $\sin\theta = \frac{-\sqrt{5}}{2}, \tan\theta = -\sqrt{5}$

32. Solve $2 \ln|x-3| - \ln|x+5| = 2 \ln 2$ for $x > 0$.

33. A hyperbola has vertices at $(0, -4)$ and $(0, 4)$ and passes through the point $(-3, -5)$. What are the coordinates of one of the foci?

(A) $(0, 4+\sqrt{2})$

(B) $(-3, 4+\sqrt{2})$

(C) $(0, 4\sqrt{2})$

(D) $(0, 4-\sqrt{2})$

(E) $(-3, 4\sqrt{2})$

34. Solve $3^{2x^2-12} = 1$ for x.

(A) $\pm\sqrt{6}$

(B) $\pm\dfrac{\sqrt{13}}{2}$

(C) ± 6

(D) $\pm 2\sqrt{3}$

(E) $\pm\dfrac{\sqrt{6}}{6}$

35.

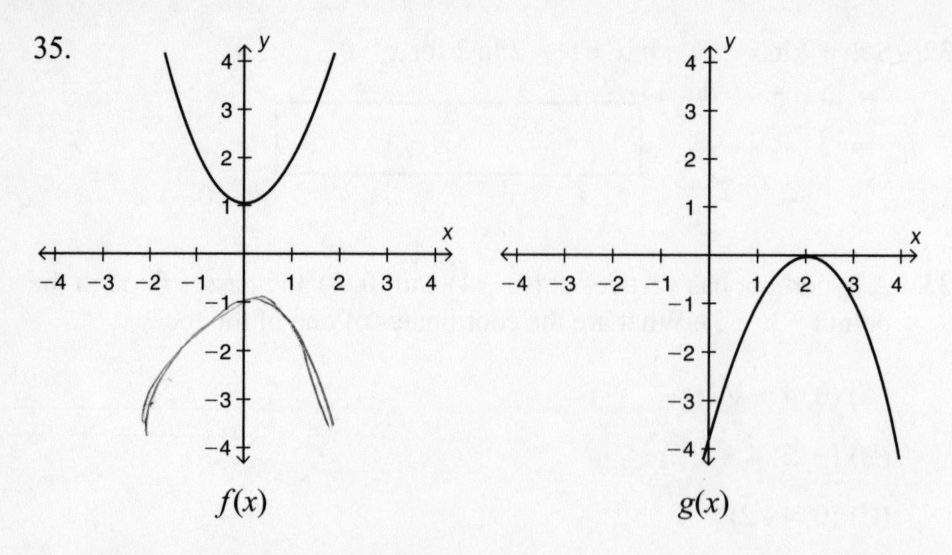

Given that $g(x)$ is a translation of $f(x)$ in the graphs above. Which function could represent $g(x)$?

(A) $g(x) = -f(x) - 1$

(B) $g(x) = f(-x) + 1$

(C) $g(x) = -f(x+2) + 1$

(D) $g(x) = -f(x-2) + 1$

(E) $g(x) = -f(x-2) - 1$

36. Solve $3\cos^{-1}\left(\frac{x}{2}\right) = \pi$ for x.

37. The graph of which of the following functions has a horizontal asymptote at $y = 0$ and a vertical asymptote at $x = 0$?

(A) $y = e^x$

(B) $y = \cos x$

(C) $y = \sin x$

(D) $y = \frac{2}{x}$

(E) $y = \frac{1}{x} - 1$

38. Given $f(x) = ax^2 - 5$, $g(x) = b^{3x}$, and $h(x) = f(x)[g(x) + ab]$. If $f(1) = 2$ and $g(1) = 8$, find $h(0)$.

(A) 0 (D) 81

(B) −75 (E) 6

(C) 32

39.

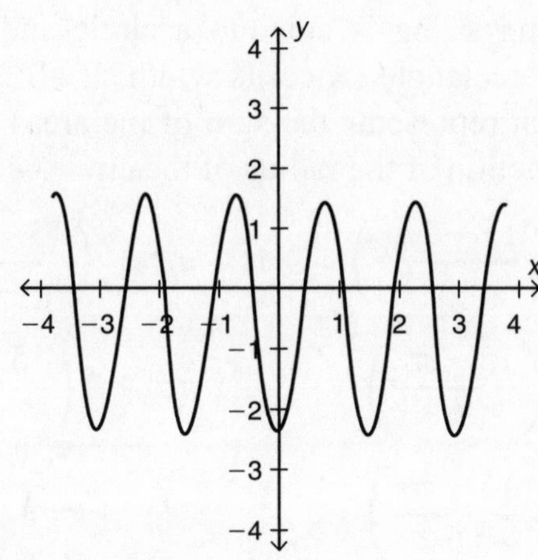

Which function represents the above graph?

(A) $y = \cos\left[3\left(x + \frac{2\pi}{3}\right)\right] + 1$

(B) $y = 2\cos\left[3\left(x - \frac{2\pi}{3}\right)\right] + 1$

(C) $y = 2\sin\left[2\left(x - \frac{\pi}{4}\right)\right] - \frac{1}{2}$

(D) $y = 2\cos\left[4\left(x + \frac{3\pi}{4}\right)\right] - \frac{1}{2}$

(E) $y = \sin\left[3\left(x + \frac{\pi}{2}\right)\right] - 1$

40. What must be true about a, b, c, and d for $f(x) = ax^3 + bx^2 + cx + d$ to be an odd function?

(A) $a = 0$

(B) $a = c = 0$

(C) $b = d = 0$

(D) only $b = 0$

(E) only $d = 0$

41. A 15-foot-long string is cut into a circle and a rectangle with the length of the rectangle twice its width. If all the string is used, find a function that represents the sum of the areas of the circle and rectangle as a function of the radius of the circle.

(A) $\pi r^2 + 2\left(\frac{15 - 2\pi r}{6}\right)^2$

(B) $\pi r^2 + 2\left(\frac{15 - \pi r}{6}\right)^2$

(C) $2\pi r^2 + \left(\frac{15 - \pi r}{6}\right)^2$

(D) $\pi r^2 + \left(\frac{15 - 2\pi r}{2}\right)^2$

(E) $\pi r^2 + 2\left(\frac{15 - \pi r}{4}\right)^2$

42. What is the period of the function $y = 2\sin(\frac{3\pi}{4}x - 1) + \frac{2}{3}$?

(A) $\frac{3}{4}$

(B) $\frac{4\pi}{3}$

(C) $\frac{8}{3}$

(D) $\frac{3}{8}$

(E) $\frac{8\pi}{3}$

43. $f(x) = 8x^3 + 4x^2 + x - 2$

Using the Rational Root Theorem, list the potential rational roots of the above polynomial.

(A) $x = \pm 2, \pm 4, \pm 8$

(B) $x = \pm 1, \pm 2, \pm 4, \pm \frac{1}{2}$

(C) $y = \pm 1, \pm 2, \pm \frac{1}{2}, \pm \frac{1}{4}, \pm \frac{1}{8}$

(D) $y = 1, 2, 4, 8, \frac{1}{2}$

(E) $x = \pm 1, \pm 2, \pm \frac{1}{2}, \pm \frac{1}{4}, \pm \frac{1}{8}$

44.

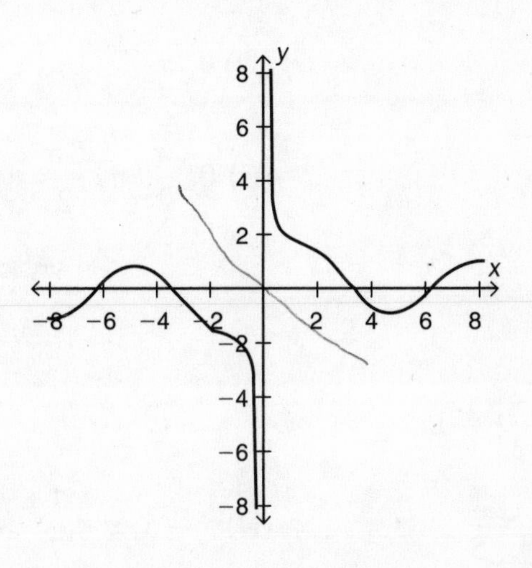

Given $f(x) = \sin x$ and $g(x) = \frac{1}{x}$. The graph above represents which of the following?

(A) $\frac{f(x)}{g(x)}$

(B) $f(x) - g(x)$

(C) $\frac{g(x)}{f(x)}$

(D) $f(x) \times g(x)$

(E) $f(x) + g(x)$

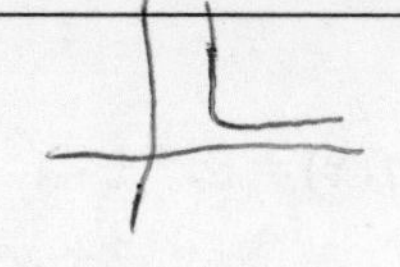

45. Which is true about the graph of $y = \frac{1}{x} + 2$?

I. Contains points only in the first and third quadrants

II. Never intersects the x-axis

III. Never intersects the y-axis

IV. Contains points only in the first quadrant

(A) I only

(B) I and III only

(C) I and II only

(D) III only

(E) I, II, and III only

46. Solve $\cos^2 x + \sin x + 1 = 0$ for x in $[0, 2\pi)$.

(A) $0, \frac{\pi}{2}, \pi$

(B) $\frac{3\pi}{2}$

(C) $\frac{\pi}{2}, \frac{3\pi}{2}$

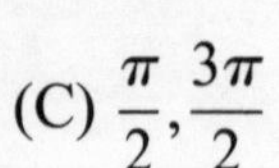

(D) $0, \pi$

(E) $0, \frac{\pi}{2}, \pi, \frac{3\pi}{2}$

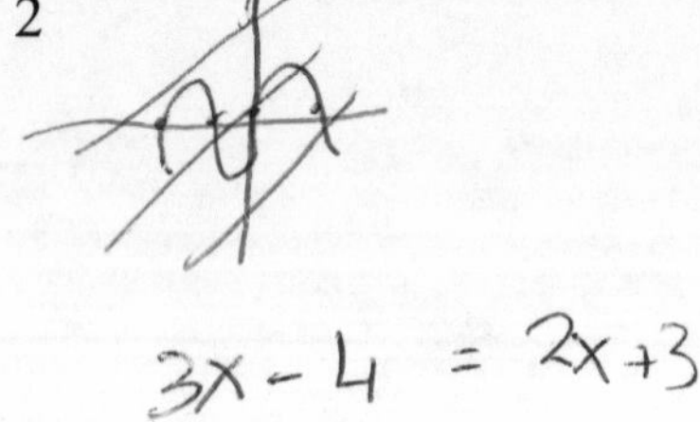

47. Solve $\left|\frac{3x-4}{2x+3}\right| = 1$.

(A) $x = 7, x = \frac{1}{5}$

(B) $x = 7, x = -\frac{1}{5}$

(C) $x = -7, x = -\frac{1}{5}$

(D) $x = 5, x = \frac{1}{7}$

(E) $x = -5, x = \frac{1}{7}$

48. What is the maximum value of $y = -2x^2 + 4x + 1$?

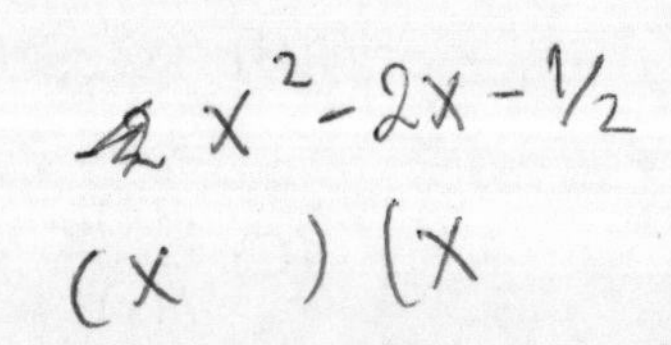

Answer Key
Practice Test 2

1. (D)
2. (A)
3. (E)
4. 5
5. (A)
6. (B)
7. (D)
8. (E)
9. (D)
10. (B)
11. 4
12. (B)
13. (B)
14. −3
15. (A)
16. (E)
17. (B)
18. (D)
19. (D)
20. (A)
21. (C)
22. (B)
23. (C)
24. (D)
25. (D)
26. 3
27. (C)
28. (A)
29. (B)
30. (B)
31. (D)
32. 11
33. (C)
34. (A)
35. (D)
36. 1
37. (D)
38. (B)
39. (D)
40. (C)
41. (A)
42. (C)
43. (E)
44. (E)
45. (D)
46. (B)
47. (A)
48. 3

Detailed Explanations of Answers

Practice Test 2

1. **(D)**

$$\frac{1}{2}\ln(3x+4)-1=\ln 2$$

We solve by putting both terms involving "ln" on the same side of the equation.

$$\frac{1}{2}\ln(3x+4)-\ln 2=1$$

We now simplify the left side of the equation by using the Rules of Logarithms.

$$\ln\sqrt{3x+4}-\ln 2=1,\text{ since } r\ln a=\ln a^r$$

$$\ln\left(\frac{\sqrt{3x+4}}{2}\right)=1,\text{ since } \ln a-\ln b=\ln\frac{a}{b}.$$

In order to solve for x, we write the equation using base e.

$$e^{\ln\left(\frac{\sqrt{3x+4}}{2}\right)}=e^1$$

This simplifies to:

$$\frac{\sqrt{3x+4}}{2}=e$$

$$\sqrt{3x+4}=2e$$

$$3x+4=4e^2$$

$$x=\frac{4e^2-4}{3}\approx 8.519$$

2. **(A)**

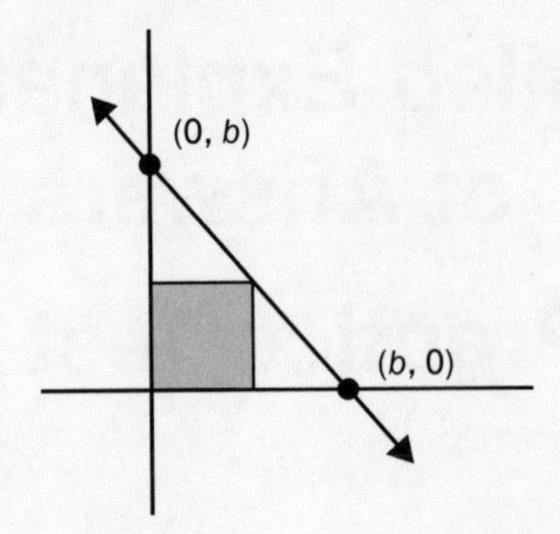

Given the line $y = -x + b$, the x- and y-intercepts are $(b, 0)$ and $(0, b)$, respectively.

Area of large triangle enclosed by line and axes: $A = \frac{1}{2} b \times b = \frac{b^2}{2}$

Area of square: c^2

Difference: $\frac{b^2}{2} - c^2$

3. **(E)**

$3\sin^2 4x + 3\sin 4x = 1$

This is a quadratic equation in sin $4x$ that does not factor. Therefore, we use the quadratic formula with $a = 3$, $b = 3$, $c = -1$.

$$\sin 4x = \frac{-3 \pm \sqrt{3^2 - 4(3)(-1)}}{2(3)} = \frac{-3 \pm \sqrt{21}}{6} \approx \frac{-3 \pm 4.583}{6}$$

$\sin 4x \approx -1.264$ $\quad$ $\sin 4x \approx 0.264$

(invalid—out of range) $\quad$ $4x = \sin^{-1}(.0264)$

$$x = \frac{\sin^{-1}(0.264)}{4} \approx .067$$

4. **(5)**

$x^2 + y^2 - 2x - 10y = 10$ and $x^2 + y^2 - 8x - 2y + 1 = 0$

We rewrite each equation into standard form to get the coordinates of the center of each circle.

First circle: $(x^2 - 2x + \underline{\quad}) + (y^2 - 10y + \underline{\quad}) = 10 + \underline{\quad} + \underline{\quad}$

$(x^2 - 2x + \underline{1}) + (y^2 - 10y + \underline{25}) = 10 + \underline{1} + \underline{25}$

$(x - 1)^2 + (y - 5)^2 = 36$

center: $(1, 5)$

Second circle: $x^2 + y^2 - 8x - 2y + 1 = 0$

$$(x^2 - 8x + \_\_\_) + (y^2 - 2y + \_\_\_) = -1 + \_\_\_ + \_\_\_$$

$$(x^2 - 8x + \underline{16}) + (y^2 - 2y + \underline{1}) = -1 + \underline{16} + \underline{1}$$

$$(x-4)^2 + (y-1)^2 = 16$$

center: (4, 1)

We now use the distance formula to calculate the distance between the centers.

$$d = \sqrt{(1-4)^2 + (5-1)^2} = \sqrt{(-3)^2 + (4)^2} = \sqrt{25} = 5$$

5. **(A)**

We start with $y = x^2$. To shift to the right 7 units, we translate the function to $y = (x-7)^2$. To shift down one unit, we have $y = (x-7)^2 - 1$. To open downward and 4 times wider than the parent function, we multiply by $-\frac{1}{4}$.

$$y = -\frac{1}{4}(x-7)^2 - 1$$

6. **(B)**

A function is odd if $f(-x) = -f(x)$. We check each function listed.

I. For the first function, we know that x^2 is an even function, $\sin x$ is an odd function, and $\cos x$ is even. Therefore, the product of $x^2 \sin x \cos x$ is an odd function.

II. Since $\sin x$ is an odd function, then $\frac{2}{\sin x}$ is an odd function.

III. We use the test above to check $\ln|2 - x|$.

$$f(-x) = \ln|2 - (-x)| = \ln|2 + x| \neq -f(x)$$

IV. We also check $\frac{e^x}{x^2}$.

$$f(-x) = \frac{e^{(-x)}}{(-x)^2} = \frac{e^{-x}}{x^2} \neq -f(x)$$

Only I and II are odd, which is choice (B).

7. **(D)**

Town	Population	Annual Growth Rate
A	35,000	4%
B	26,000	7.5%

The population of Town A:

After 1 year: $35000 + 35000(.04) = 35000(1.04)$

After 2 years: $35000(1.04)(1.04) = 35000(1.04)^2$

After t years: $35000(1.04)^t$

Similarly, the population of Town B after t years is: $26000(1.075)^t$. We are interested in knowing when these two populations are equal.

$$35000(1.04)^t = 26000(1.075)^t$$

$$\frac{35000}{26000} = \frac{(1.075)^t}{(1.04)^t} = \left(\frac{1.075}{1.04}\right)^t$$

$$1.35 = (1.034)^t$$

$$\ln(1.35) = \ln(1.034)^t$$

$$\ln(1.35) = t\ln(1.034)$$

$$\frac{\ln(1.35)}{\ln(1.034)} = t$$

$$t \approx 9 \text{ years}$$

8. **(E)**

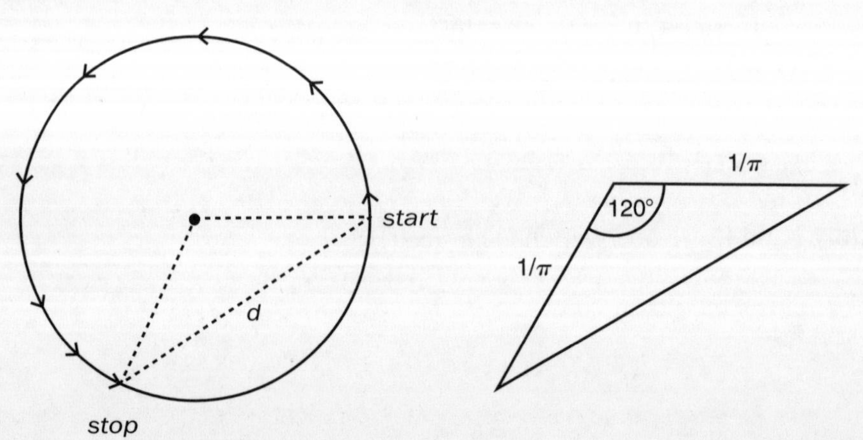

Since you walked $\frac{2}{3}$ around the circle, the angle you walked around is $\frac{2}{3} \times 360° = 240°$. Since the circumference is 2 miles (given) and $c = 2\pi r$, we can solve for r.

$$2 = 2\pi r$$

$$r = \frac{1}{\pi}$$

We can use the Law of Cosines to find the missing side, which is a chord of the circle. The Law of Cosines can be written as:

$$d^2 = a^2 + b^2 - 2ab\cos D$$

$$d^2 = \left(\frac{1}{\pi}\right)^2 + \left(\frac{1}{\pi}\right)^2 - 2\left(\frac{1}{\pi}\right)\left(\frac{1}{\pi}\right)\cos(120)°$$

$$d^2 = \left(\frac{1}{\pi}\right)^2 + \left(\frac{1}{\pi}\right)^2 - 2\left(\frac{1}{\pi}\right)\left(\frac{1}{\pi}\right)\left(-\frac{1}{2}\right)$$

$$d^2 = \left(\frac{1}{\pi}\right)^2 + \left(\frac{1}{\pi}\right)^2 + \left(\frac{1}{\pi}\right)^2 = \frac{3}{\pi^2}$$

$$d = \frac{\sqrt{3}}{\pi} \approx .55 \text{ miles}$$

9. **(D)**

This problem is best worked by elimination of the choices. From their graphs, we can see that the graph of the product will be negative for $x < 0$, since one function is negative and the other is positive for $x < 0$. Hence $h(x) = e^x$ (choice (B)) can be eliminated since it is positive for $x < 0$. Since we see that the functions are each equal to 0 at $x = 0$, their product at $x = 0$ will be equal to 0. This means that (0, 0) is on the graph. Hence, $h(x) = \frac{1}{x}$ (choice (A)) and $h(x) = \frac{e^x}{x}$ (choice (E)) can be eliminated since the functions are not defined at $x = 0$. We now consider choice (C): $h(x) = xe^2$. This is a line with slope e^2, which is a constant. By examining f and g at $x = 1$, we see that their product would be less than 1, since f at $x = 1$ is about $\frac{1}{3}$ and the g at $x = 1$ appears to be equal to 1. This would then eliminate choice (C), since its value at $x = 1$ would be e^2, which is much greater than 1. So, by elimination, the only possible answer from the given choices is (D). Choice (D) also fits for $x < 0$, $x = 0$, and $x > 0$.

10. **(B)**

When the vertices of a hyperbola are located on the y-axis, the general equation is given by $\frac{y^2}{a^2} - \frac{x^2}{b^2} = 1$. Then $y = \pm\frac{a}{b}x$ are the equations of the asymptotes. We can see that the center is at (0, 0), since the asymptotes go through the center. Since $a = 3$ and $b = 4$, we can set up the equation as $\frac{y^2}{3^2} - \frac{x^2}{4^2} = 1$, which becomes $\frac{y^2}{9} - \frac{x^2}{16} = 1$. Multiplying by 144, we get $16y^2 - 9x^2 = 144$.

11. **(4)**

x	$f(x)$	$g(x)$	$f \circ g$
0	0	3	5
1	2	2	?
2	4	1	2
3	5	0	0

The missing element in the table is $f(g(1))$. We first find $g(1)$ from the table, which is 2. We now find $f(2)$, which is 4.

12. **(B)**

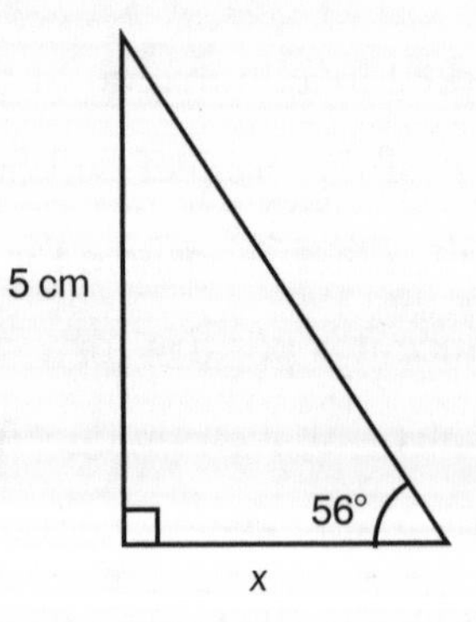

$$\tan 56° = \frac{5}{x}$$

$$x = \frac{5}{\tan 56°} \approx 3.373 \text{ cm}$$

13. **(B)**

We put the ellipse into standard form:

$$25x^2 + 4y^2 + 100x + 16y = -16$$

$$25(x^2 + 4x + \underline{\quad}) + 4(y^2 + 4y + \underline{\quad}) = -16 + \underline{\quad} + \underline{\quad}$$

$$25(x^2 + 4x + \underline{\ 4\ }) + 4(y^2 + 4y + \underline{\ 4\ }) = -16 + \underline{100} + \underline{\ 16\ }$$

$$25(x + 2)^2 + 4(y + 2)^2 = 100$$

$$\frac{(x + 2)^2}{4} + \frac{(y + 2)^2}{25} = 1$$

The center is $(-2, -2)$, $a = \pm 2$, and $b = \pm 5$. Since $c^2 = b^2 - a^2$, then $c = \pm\sqrt{21}$. The major axis is the y-axis, so we move $\pm\sqrt{21}$ units from the y-coordinate of the center. Therefore, the coordinates of one of the foci is $(-2, -2 + \sqrt{21})$.

14. **(–3)**

$$f(x) = ab^x$$

Since $f(1) = -6$, we have $f(1) = ab^1 = ab = -6$. Since $f(-1) = -\frac{3}{2}$, we have

$$f(-1) = ab^{-1} = \frac{a}{b} = -\frac{3}{2}.$$

Our two equations are:

$$ab = -6$$

$$\frac{a}{b} = -\frac{3}{2}$$

Since $b > 0$, then a must be negative. From the second equation, we can assign $a = -3$ and $b = 2$.

15. **(A)**

$$P(t) = 750 \sin \frac{\pi}{6}t + 800$$

$$1500 = 750 \sin \frac{\pi}{6}t + 800$$

$$700 = 750 \sin \frac{\pi}{6}t$$

$$\frac{70}{75} = \sin \frac{\pi}{6}t$$

$$.93 = \sin \frac{\pi}{6}t$$

$$\sin^{-1}(.93) = \frac{\pi}{6}t$$

$$t = \frac{6 \sin^{-1}(.93)}{\pi} \approx 2.3, \text{ or approximately 2 months, 10 days.}$$

We add this to January 1, 2008, and get March 10, 2008.

16. **(E)**

The two graphs intersect when $x^2 = mx + b$ or $x^2 - mx - b = 0$. Using the quadratic formula to solve, we get

$$x = \frac{m \pm \sqrt{(-m)^2 - 4(1)(-b)}}{2(1)} = \frac{m \pm \sqrt{m^2 + 4b}}{2}$$

If they do not intersect, then this solution is complex, which means the discriminant is less than zero, or $m^2 + 4b < 0$. This also indicates that $b < 0$, since this is the only way that $m^2 + 4b$ could be less than 0.

17. **(B)** $p = \dfrac{\frac{lr}{12}}{1 - (1 + \frac{r}{12})^{-m}}$, where p is the monthly payment, l is the loan amount, r is the annual interest rate (APR), and m is the number of months of the loan.

For the 8.9% loan payment:

$$p = \frac{\frac{(20000)(.089)}{12}}{1 - \left(1 + \frac{.089}{12}\right)^{-60}} \approx 414.20$$

For the 5.6% loan payment

$$p=\frac{\frac{(20000)(.056)}{12}}{1-\left(1+\frac{.056}{12}\right)^{-60}}\approx 382.95$$

Difference: \$31.25

18. **(D)**

$$N=N_0e^{kt}$$

If the bacteria doubles every 8 hours, then $\frac{N}{N_0}=\frac{2}{1}$. This information allows us to find k. Using the formula $N=N_0e^{kt}$ or $\frac{N}{N_0}=e^{kt}$ and substituting we get:

$$\frac{2}{1}=e^{k(8)} \quad\Rightarrow\quad 2=e^{8k}$$

We solve for k by writing the equation with logarithms.

$$\ln 2=\ln e^{8k} \quad\Rightarrow\quad \ln 2=8k \quad\Rightarrow\quad k=\frac{\ln 2}{8}\approx .08664$$

We now can estimate N after 30 hours.

$$N=N_0e^{(.08664)(30)}\approx N_0\times(13.45)$$

19. **(D)**

Cammy's rate: 3 ft/sec + 2.5 ft/sec = 5.5 ft/sec

Since $d=r\times t$, we have $t=\frac{d}{r}$. Cammy will walk 100 feet at 5.5 ft/sec.

$$t=\frac{d}{r}=\frac{100}{5.5}\approx 18.18 \text{ seconds}$$

Abby's rate: Abby must walk 85 feet in 18.18 seconds, so we can find her rate.

$$r=\frac{d}{t}=\frac{85}{18.18}\approx 4.7 \text{ ft/sec}$$

20. **(A)**

$$f(x) = \frac{x^2 + x - 2}{3x^2 - x - 10}$$

For the horizontal asymptote of rational functions, we only have to consider the coefficients of the highest degree in the numerator and the denominator. Since they have the same degree, the equation of the horizontal asymptote will be the ratio of their coefficients, i.e., $y = \frac{1}{3}$.

21. **(C)**

$$f(x) = e^x - x^2 \text{ and } g(x) = \sqrt{x - 2}$$

To find $g(f(2))$, we first find $f(2)$.

Since $f(x) = e^x - x^2, f(2) = e^2 - 2^2 = e^2 - 4$. We now find

$$g(f(2)) = g(e^2 - 4) = \sqrt{(e^2 - 4) - 2} = \sqrt{e^2 - 6}$$

(We can verify that $\sqrt{e^2 - 6}$ is nonnegative so that $e^2 - 4$ is in the domain of $g(x)$.)

22. **(B)**

If the graph passes through only two quadrants, then it cannot cross the x-axis. The graph is completely above the x-axis or has a vertex on the x-axis. It cannot be below the x-axis since a is positive ($a = 1$ given). Therefore, there are either complex solutions (and the graph is completely above the x-axis) or there is a double root solution. For complex solutions, $b^2 - 4ac < 0$, and for a double root solution, $b^2 - 4ac = 0$. Therefore, the solution is $b^2 - 4ac \leq 0$, or in this problem since $a = 1$, $b^2 - 4c \leq 0$, which means $c \geq \frac{b^2}{4}$.

23. **(C)**

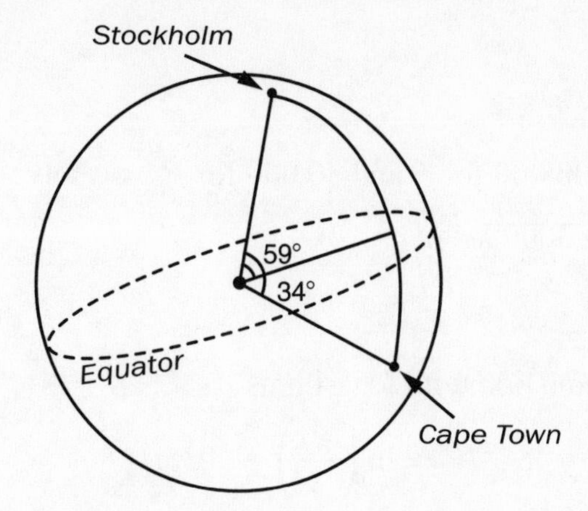

We can determine the distance by using the formula for arc length: $s = r\theta$, where r is the radius of Earth (approximately 3900 miles). We can determine the angle by using the latitude values. Since one city is at latitude 59°N and the other is at 34°S, the angle between them is $59° + 34° = 93°$. We convert the degrees into a radian measure: $\frac{93(\pi)}{180}$.

Therefore, $s = \frac{93(\pi)}{180}(3900) \approx 6330$ miles.

24. **(D)**

$$f(x) = \frac{x}{2x-1}$$

To form the inverse, we first write the equation as

$$y = \frac{x}{2x-1}$$

We now switch the x and y and solve for the new y.

$$x = \frac{y}{2y-1}$$

$$x(2y-1) = y$$

$$2xy - x = y$$

$$2xy - y = x$$

$$y(2x-1) = x$$

$$y = \frac{x}{2x-1}$$

25. **(D)**

$$P = P_0 e^{.035t}$$

Substituting 800,000 for P and 30,000 for P_0, we solve for t.

$$800000 = 30000e^{.035t}$$

$$\frac{800000}{30000} = e^{.035t} \Rightarrow \frac{80}{3} = e^{.035t}$$

We write the equation with logarithms:

$$\ln\left(\frac{80}{3}\right) = \ln e^{.035t} \Rightarrow \ln\left(\frac{80}{3}\right) = .035t$$

Solving for t:

$$t = \frac{\ln\left(\frac{80}{3}\right)}{.035} \approx 93.8 \text{ hours}$$

26. **(3)**

$$\log_{\sqrt{5}} 125 = 2x$$

We rewrite the problem as an exponential equation:

$$\left(\sqrt{5}\right)^{2x} = 125$$

We now write both sides of the equation with base 5.

$$\left[(5)^{\frac{1}{2}}\right]^{2x} = 5^3 \Rightarrow 5^x = 5^3 \Rightarrow x = 3$$

27. **(C)**

$$\frac{\sin\theta}{1 + \cos\theta} + \frac{1 + \cos\theta}{\sin\theta}$$

Finding a common denominator so we can combine, we get

$$\frac{\sin\theta(\sin\theta)}{(1 + \cos\theta)(\sin\theta)} + \frac{(1 + \cos\theta)(1 + \cos\theta)}{\sin\theta(1 + \cos\theta)}$$

$$\frac{\sin^2\theta + (1 + \cos\theta)^2}{(1 + \cos\theta)\sin\theta}$$

$$\frac{\sin^2\theta + (1 + 2\cos\theta + \cos^2\theta)}{(1 + \cos\theta)\sin\theta}$$

Rearranging terms:

$$\frac{\sin^2\theta + \cos^2\theta + 1 + 2\cos\theta}{(1+\cos\theta)\sin\theta}$$

Since $\sin^2\theta + \cos^2\theta = 1$, we have

$$\frac{2 + 2\cos\theta}{(1+\cos\theta)\sin\theta}.$$

Factoring the numerator and reducing,

$$\frac{2(1+\cos\theta)}{(1+\cos\theta)\sin\theta} =$$

$\dfrac{2}{\sin\theta}$, which is choice (B).

Choice (A) is equivalent to this answer since

$$2\csc\theta = \frac{2}{\sin\theta}.$$

Choice (D) is equivalent to this answer since

$$\frac{2\tan\theta}{\sin^2\theta\sec\theta} = \frac{2\dfrac{\sin\theta}{\cos\theta}}{\sin^2\theta\left(\dfrac{1}{\cos\theta}\right)} = \frac{2}{\sin\theta}.$$

Choice (E) is equivalent to this answer since

$$\frac{2\sec\theta}{\tan\theta} = \frac{2\left(\dfrac{1}{\cos\theta}\right)}{\left(\dfrac{\sin\theta}{\cos\theta}\right)} = \frac{2}{\sin\theta}.$$

We can show choice (C) is not equivalent:

$$\frac{2\sin\theta}{\tan\theta} = \frac{2\sin\theta}{\left(\dfrac{\sin\theta}{\cos\theta}\right)} = 2\cos\theta \neq \frac{2}{\sin\theta}$$

28. **(A)**

$$y = \frac{1}{2}\sin 3x + \frac{3}{2}$$

Since the amplitude is $\frac{1}{2}$ and $y = \frac{3}{2}$ is the line through the middle of the graph, we only need to go $\frac{1}{2}$ units below the middle of the graph to get the minimum value of the function. Therefore, the minimum value is $y = 1$.

29. **(B)**

$$y = 2\sin^{-1}\left(\frac{x}{2}\right)$$

The domain of the inverse sine function is $[-1, 1]$. Therefore, the domain of $\sin^{-1}\left(\frac{x}{2}\right)$ is $-1 \le \frac{x}{2} \le 1$ or, solving for x, $-2 \le x \le 2$.

30. **(B)**

$$x^3 + 4x - 23x + 6$$

We can find a factor of a polynomial by several methods. One way is to take each $(x - a)$ binomial from our choices and evaluate the function at each $x = a$. If $f(a) = 0$, then $x - a$ is a factor.

We check $a = 1$ in $(x - 1)$: $f(1) = (1)^3 + 4(1)^2 - 23(1) + 6 = -12 \neq 0$.

Therefore, $(x - 1)$ is *not* a factor.

We check $a = 3$ in $(x - 3)$: $f(3) = (3)^3 + 4(3)^2 - 23(3) + 6 = 0$.

Therefore, $(x - 3)$ is a factor.

We check $a = -1$ in $(x + 1)$: $f(-1) = (-1)^3 + 4(-1)^2 - 23(-1) + 6 = 32 \neq 0$.

Therefore, $(x + 1)$ is *not* a factor.

We check $a = 2$ in $(x - 2)$: $f(2) = (2)^3 + 4(2)^2 - 23(2) + 6 = -24 \neq 0$.

Therefore, $(x - 2)$ is *not* a factor.

We check $a = -2$ in $(x + 2)$: $f(-2) = (-2)^3 + 4(-2)^2 - 23(-2) + 6 = 60 \neq 0$.

Therefore, $(x + 2)$ is *not* a factor.

The only one of our choices that is a factor is $(x - 3)$.

31. **(D)**

$$\cos\theta = \frac{2}{3}$$

Since $\cos\theta = \frac{x}{r}$ and we are given $\cos\theta = \frac{2}{3}$, then we know $\frac{x}{r} = \frac{2}{3}$. This allows us to find y since $x^2 + y^2 = r^2$. Substituting these values we get $(2)^2 + y^2 = (3)^2$, or $y = \pm\sqrt{5}$. Since we are given that the angle is in the fourth quadrant, then $y = -\sqrt{5}$. We can now find $\sin\theta = \frac{-\sqrt{5}}{3}$ and $\tan\theta = \frac{-\sqrt{5}}{2}$.

32. **(11)**

$$2\ln|x-3| - \ln|x+5| = 2\ln 2$$

We use the Rules of Logarithms to rewrite the equation:

$$\ln|x-3|^2 - \ln|x+5| = \ln 2^2$$

$$\ln\left|\frac{(x-3)^2}{x+5}\right| = \ln 4$$

$$\frac{(x-3)^2}{x+5} = 4 \qquad \text{(since } x > 0,\ |x+5| = x+5\text{)}$$

$$(x^2 - 6x + 9) = 4(x+5)$$

$$x^2 - 10x - 11 = 0$$

$$(x-11)(x+1) = 0$$

$$x - 11 = 0 \qquad x + 1 = 0$$

$$x = 11 \qquad x = -1$$ (We discard this answer since we were given $x > 0$.)

33. **(C)**

We know that the standard form of this hyperbola will be $\frac{y^2}{16} - \frac{x^2}{b^2} = 1$ since the graph crosses the y-axis and has vertices at $(0, -4)$ and $(0, 4)$. Since the center of the hyperbola is the midpoint between the vertices, then the coordinates of the center are $(0, 0)$. The point $(-3, -5)$ will allow us to find b, since this value of x and y must satisfy the equation.

$$\frac{(-5)^2}{16} - \frac{(-3)^2}{b^2} = 1$$

$$\frac{25}{16} - \frac{9}{b^2} = 1$$

We now multiply both sides by $16b^2$ to clear out fractions.

$$25b^2 - 144 = 16b^2$$

$$9b^2 = 144$$

$$b^2 = 16, \text{ or } b = \pm 4$$

Therefore, the equation is $\frac{y^2}{16} - \frac{x^2}{16} = 1$ or $y^2 - x^2 = 16$. We can use the values of a and b to find c, which will allow us to find the foci. Since $c^2 = a^2 + b^2$, we have $c^2 = 16 + 16 = 32$. Then $c = \pm\sqrt{32} = 4\sqrt{2}$. The foci will be c units from the center of the hyperbola. Hence, $(0, 4\sqrt{2})$ and $(0, -4\sqrt{2})$ are the coordinates of the foci.

34. **(A)**

$$3^{2x^2 - 12} = 1$$

To solve the equation we need to write the number 1 with base 3, or as 3°. Therefore, we have

$$3^{2x^2 - 12} = 3^{\circ}$$

Since the bases are equal, we can equate the exponents.

$$2x^2 - 12 = 0 \quad \Rightarrow \quad x^2 = 6 \quad \Rightarrow \quad x = \pm\sqrt{6}$$

35. **(D)**

The vertex has been translated from (0, 1) to (2, 0). Therefore, the graph is one unit down and two units to the right, and opening downward.

We first go down one unit, so we will have $f(x) - 1$.

We now move the vertex two units to the right: $f(x - 2) - 1$.

We now have the function open downward: $-[f(x - 2) - 1]$, or

$-f(x - 2) + 1$

36. **(1)**

$$3\cos^{-1}\left(\frac{x}{2}\right) = \pi$$

To solve this problem, we first divide by 3.

$$\cos^{-1}\left(\frac{x}{2}\right) = \frac{\pi}{3}$$

We now change from an inverse equation:

$$\cos\left(\frac{\pi}{3}\right) = \frac{x}{2}$$

Since we know that $\cos\left(\frac{\pi}{3}\right) = \frac{1}{2}$, then $x = 1$.

37. **(D)**

We can begin by eliminating choices.

$y = e^x$ does not have a vertical asymptote.

$y = \cos x$ does not have a vertical or horizontal asymptote.

$y = \sin x$ does not have a vertical or horizontal asymptote.

$y = \frac{2}{x}$ has a vertical asymptote at $x = 0$ and a horizontal asymptote at $y = 0$.

$y = \frac{1}{x} - 1$ has a vertical asymptote at $x = 0$ and a horizontal asymptote at $y = -1$.

38. **(B)**

$$f(x) = ax^2 - 5,\ g(x) = b^{3x},\ \text{and}\ h(x) = f(x)\,[g(x) + ab]$$

Given $f(x) = ax^2 - 5$ and $f(1) = 2$. Then

$$f(1) = a(1)^2 - 5 = 2 \quad \Rightarrow \quad a = 7$$

Given $g(x) = b^{3x}$ and $g(1) = 8$. Then

$$g(1) = b^{3(1)} = b^3 = 8 \quad \Rightarrow \quad b = 2$$

Given $h(x) = f(x)\,[g(x) + ab]$

Substituting the values of a and b into f and g, we get

$$h(x) = (7x^2 - 5)[2^{3x} + (7)(2)]$$

$$h(0) = (7(0)^2 - 5)[2^0 + 14] = (-5)(15) = -75$$

39. **(D)**

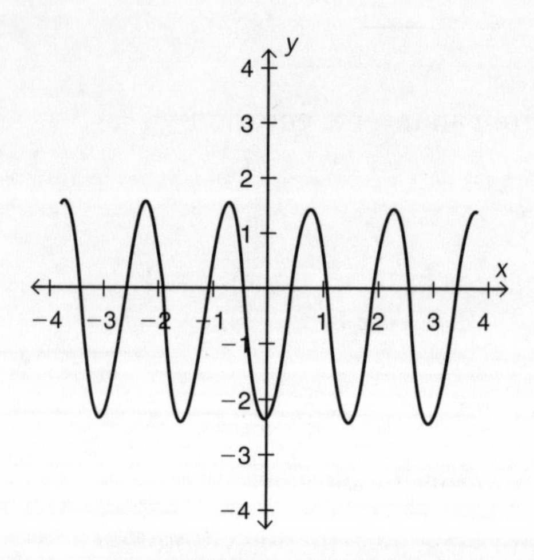

From the graph, we try to determine as many values as possible for the general form of sinusoidal equations: $y = A\ \sin(B(\theta - h)) + k$ or $y = A\ \cos(B(\theta - h)) + k$, where $|A|$ is the amplitude, h is the phase or horizontal shift, $y = k$ is the horizontal line through the middle of the graph, and $|B|$ is the number of cycles completed within 2π radians. From this graph, we see that the difference between the maximum and minimum values is 4 units, which is also twice the amplitude. Therefore, $A = 2$. Since $y = k$ is the equation of the line through the middle of the graph, we can go 2 units from the top or bottom to see that $y = -\frac{1}{2}$ is the line through the middle of the graph, so $k = -\frac{1}{2}$. From the graph we can also see that there are four complete cycles from approximately $-\pi$ to π. So, B is equal to 4. The only graph that fits $A = 2$, $k = -\frac{1}{2}$, and $B = 4$ is graph (D).

40. **(C)**

$$f(x) = ax^3 + bx^2 + cx + d$$

An odd function has $f(-x) = -f(x)$. For $f(x) = ax^3 + bx^2 + cx + d$ to be an odd function, we must have only odd powers. Therefore, the coefficients of all even powers and the constant must equal 0.

41. **(A)**

The circle will use $2\pi r$ feet of the string. The rectangle will use $2w + 2l = 2(w) + 2(2w) = 6w$, since the length is twice the width. Since the circle and rectangle will use the entire length of the string, we have

$$2\pi r + 6w = 15$$

We can use this to solve for w in terms of r.

$$w = \frac{15 - 2\pi r}{6}$$

We define the areas:

$$A_{circle} = \pi r^2$$

$$A_{rectangle} = l \times w = (2w) \times w = 2w^2 = 2\left(\frac{15 - 2\pi r}{6}\right)^2$$

The sum of the areas will be:

$$A_{total} = A_{circle} + A_{rectangle}$$

$$A_{total} = \pi r^2 + 2\left(\frac{15 - 2\pi r}{6}\right)^2$$

42. **(C)**

$$y = 2\sin\left(\frac{3\pi}{4}x - 1\right) + \frac{2}{3}$$

In general, for $y = A \sin B(x - h) + k$, the period is $\frac{2\pi}{|B|}$. We can write the given equation in this form: $y = 2\sin\frac{3\pi}{4}\left(x - \frac{4}{3\pi}\right) + \frac{2}{3}$.

Therefore, we have $\frac{2\pi}{|B|} = \frac{2\pi}{\left(\frac{3\pi}{4}\right)} = \frac{8}{3}$.

43. **(E)**

$$f(x) = 8x^3 + 4x^2 + x - 2$$

From the Rational Root Theorem, we know that any rational roots to a polynomial must be of the form $\pm\frac{p}{q}$, where p is a factor of a_0 (the constant) and q is a factor of a_n (the coefficient of the highest power). In our problem, $p = -2$ and $q = 8$. We now list the factors of p and q.

Factors of 2: $\pm 1, \pm 2$

Factors of 8: $\pm 1, \pm 2, \pm 4, \pm 8$

Therefore, any rational solutions to the equation $8x^3 + 4x^2 + x - 2 = 0$ will be:

$$x = \pm\frac{1}{1}, \pm\frac{1}{2}, \pm\frac{1}{4}, \pm\frac{1}{8}, \pm\frac{2}{1}, \pm\frac{2}{2}, \pm\frac{2}{4}, \pm\frac{2}{8}$$

Eliminating duplicates and reducing we get:

$$x = \pm 1, \pm 2, \pm\frac{1}{2}, \pm\frac{1}{4}, \pm\frac{1}{8}$$

44. **(E)**

$$f(x) = \sin x \text{ and } g(x) = \frac{1}{x}$$

We can eliminate choices (A), (C), and (D) since these are even functions and our graph is odd. (They are even functions because f and g are both odd and the products and quotients of two odd functions are even.) We now look at choice (B). If $x < 0$, but close to 0, then $\sin x$ is negative and $\frac{1}{x}$ is negative, but $\frac{1}{x}$ is a very "large" negative number, much greater than

$\sin x$. Therefore, $\sin x - \frac{1}{x}$ will be positive, not negative, to the left of 0 and close to 0. The given graph has a "large" negative number when x is to the immediate left of 0. $f(x) + g(x)$ will give us a "large" negative number there. Choice (E) is the only choice that fits the graph.

45. **(D)**

$$y = \frac{1}{x} + 2$$

We consider the parent function $y = \frac{1}{x}$, which is in the first and third quadrants. $y = \frac{1}{x} + 2$ moves the parent function two units up. Therefore, some of the points in the third quadrant are moved up into the second quadrant. This eliminates I and IV as being true. If $y = 0$, then $x = -\frac{1}{2}$ is an intercept. This indicates that II is false. Since $x \neq 0$, there is no y-intercept, which means III is true. Therefore, the only statement that is true is statement III, which is choice (D).

46. **(B)**

$$\cos^2 x + \sin x + 1 = 0$$

We rewrite $\cos^2 x$ as $1 - \sin^2 x$ and substitute into the equation.

$$(1 - \sin^2 x) + \sin x + 1 = 0$$

Multiplying both sides by (-1) and combining like terms, we get

$$\sin^2 x - \sin x - 2 = 0$$

We now factor this quadratic and solve:

$$(\sin x + 1)(\sin x - 2) = 0$$

$$\sin x + 1 = 0 \qquad \sin x - 2 = 0$$

$$\sin x = -1 \qquad \sin x \neq 2$$

$$x = \frac{3\pi}{2}$$

47. **(A)**

$$\left|\frac{3x-4}{2x+3}\right| = 1$$

Case 1:

$$\frac{3x-4}{2x+3} = 1$$

$$3x - 4 = 2x + 3$$

$$x = 7$$

Case 2:

$$\frac{3x-4}{2x+3} = -1$$

$$3x - 4 = -2x - 3$$

$$5x = 1$$

$$x = \frac{1}{5}$$

48. **(3)**

$$y = -2x^2 + 4x + 1$$

To get the maximum value, we find the coordinates of the vertex:

$$y = -2(x^2 - 2x + \underline{\quad}) + 1 + \underline{\quad}$$

$$y = -2(x^2 - 2x + \underline{\ 1\ }) + 1 + \underline{\ 2\ }$$

$$y = -2(x-1)^2 + 3$$

The vertex is at (1, 3), and the maximum value of the parabola opening down is the y-value of the vertex, $y = 3$.

ANSWER SHEETS

CLEP Precalculus Practice Test 1

1. Ⓐ Ⓑ Ⓒ Ⓓ Ⓔ
2. Ⓐ Ⓑ Ⓒ Ⓓ Ⓔ
3. Ⓐ Ⓑ Ⓒ Ⓓ Ⓔ
4. Ⓐ Ⓑ Ⓒ Ⓓ Ⓔ
5. ▭
6. Ⓐ Ⓑ Ⓒ Ⓓ Ⓔ
7. Ⓐ Ⓑ Ⓒ Ⓓ Ⓔ
8. Ⓐ Ⓑ Ⓒ Ⓓ Ⓔ
9. Ⓐ Ⓑ Ⓒ Ⓓ Ⓔ
10. ▭
11. Ⓐ Ⓑ Ⓒ Ⓓ Ⓔ
12. Ⓐ Ⓑ Ⓒ Ⓓ Ⓔ
13. Ⓐ Ⓑ Ⓒ Ⓓ Ⓔ
14. Ⓐ Ⓑ Ⓒ Ⓓ Ⓔ
15. Ⓐ Ⓑ Ⓒ Ⓓ Ⓔ
16. Ⓐ Ⓑ Ⓒ Ⓓ Ⓔ
17. Ⓐ Ⓑ Ⓒ Ⓓ Ⓔ
18. Ⓐ Ⓑ Ⓒ Ⓓ Ⓔ
19. Ⓐ Ⓑ Ⓒ Ⓓ Ⓔ
20. Ⓐ Ⓑ Ⓒ Ⓓ Ⓔ
21. Ⓐ Ⓑ Ⓒ Ⓓ Ⓔ
22. Ⓐ Ⓑ Ⓒ Ⓓ Ⓔ
23. Ⓐ Ⓑ Ⓒ Ⓓ Ⓔ
24. Ⓐ Ⓑ Ⓒ Ⓓ Ⓔ
25. Ⓐ Ⓑ Ⓒ Ⓓ Ⓔ
26. Ⓐ Ⓑ Ⓒ Ⓓ Ⓔ
27. ▭
28. Ⓐ Ⓑ Ⓒ Ⓓ Ⓔ
29. Ⓐ Ⓑ Ⓒ Ⓓ Ⓔ
30. Ⓐ Ⓑ Ⓒ Ⓓ Ⓔ
31. Ⓐ Ⓑ Ⓒ Ⓓ Ⓔ
32. Ⓐ Ⓑ Ⓒ Ⓓ Ⓔ
33. Ⓐ Ⓑ Ⓒ Ⓓ Ⓔ
34. Ⓐ Ⓑ Ⓒ Ⓓ Ⓔ
35. Ⓐ Ⓑ Ⓒ Ⓓ Ⓔ
36. ▭
37. Ⓐ Ⓑ Ⓒ Ⓓ Ⓔ
38. Ⓐ Ⓑ Ⓒ Ⓓ Ⓔ
39. Ⓐ Ⓑ Ⓒ Ⓓ Ⓔ
40. ▭
41. Ⓐ Ⓑ Ⓒ Ⓓ Ⓔ
42. Ⓐ Ⓑ Ⓒ Ⓓ Ⓔ
43. Ⓐ Ⓑ Ⓒ Ⓓ Ⓔ
44. Ⓐ Ⓑ Ⓒ Ⓓ Ⓔ
45. ▭
46. Ⓐ Ⓑ Ⓒ Ⓓ Ⓔ
47. Ⓐ Ⓑ Ⓒ Ⓓ Ⓔ
48. ▭

CLEP Precalculus Practice Test 2

1. Ⓐ Ⓑ Ⓒ Ⓓ Ⓔ	17. Ⓐ Ⓑ Ⓒ Ⓓ Ⓔ	33. Ⓐ Ⓑ Ⓒ Ⓓ Ⓔ
2. Ⓐ Ⓑ Ⓒ Ⓓ Ⓔ	18. Ⓐ Ⓑ Ⓒ Ⓓ Ⓔ	34. Ⓐ Ⓑ Ⓒ Ⓓ Ⓔ
3. Ⓐ Ⓑ Ⓒ Ⓓ Ⓔ	19. Ⓐ Ⓑ Ⓒ Ⓓ Ⓔ	35. Ⓐ Ⓑ Ⓒ Ⓓ Ⓔ
4. ☐	20. Ⓐ Ⓑ Ⓒ Ⓓ Ⓔ	36. ☐
5. Ⓐ Ⓑ Ⓒ Ⓓ Ⓔ	21. Ⓐ Ⓑ Ⓒ Ⓓ Ⓔ	37. Ⓐ Ⓑ Ⓒ Ⓓ Ⓔ
6. Ⓐ Ⓑ Ⓒ Ⓓ Ⓔ	22. Ⓐ Ⓑ Ⓒ Ⓓ Ⓔ	38. Ⓐ Ⓑ Ⓒ Ⓓ Ⓔ
7. Ⓐ Ⓑ Ⓒ Ⓓ Ⓔ	23. Ⓐ Ⓑ Ⓒ Ⓓ Ⓔ	39. Ⓐ Ⓑ Ⓒ Ⓓ Ⓔ
8. Ⓐ Ⓑ Ⓒ Ⓓ Ⓔ	24. Ⓐ Ⓑ Ⓒ Ⓓ Ⓔ	40. Ⓐ Ⓑ Ⓒ Ⓓ Ⓔ
9. Ⓐ Ⓑ Ⓒ Ⓓ Ⓔ	25. Ⓐ Ⓑ Ⓒ Ⓓ Ⓔ	41. Ⓐ Ⓑ Ⓒ Ⓓ Ⓔ
10. Ⓐ Ⓑ Ⓒ Ⓓ Ⓔ	26. ☐	42. Ⓐ Ⓑ Ⓒ Ⓓ Ⓔ
11. ☐	27. Ⓐ Ⓑ Ⓒ Ⓓ Ⓔ	43. Ⓐ Ⓑ Ⓒ Ⓓ Ⓔ
12. Ⓐ Ⓑ Ⓒ Ⓓ Ⓔ	28. Ⓐ Ⓑ Ⓒ Ⓓ Ⓔ	44. Ⓐ Ⓑ Ⓒ Ⓓ Ⓔ
13. Ⓐ Ⓑ Ⓒ Ⓓ Ⓔ	29. Ⓐ Ⓑ Ⓒ Ⓓ Ⓔ	45. Ⓐ Ⓑ Ⓒ Ⓓ Ⓔ
14. ☐	30. Ⓐ Ⓑ Ⓒ Ⓓ Ⓔ	46. Ⓐ Ⓑ Ⓒ Ⓓ Ⓔ
15. Ⓐ Ⓑ Ⓒ Ⓓ Ⓔ	31. Ⓐ Ⓑ Ⓒ Ⓓ Ⓔ	47. Ⓐ Ⓑ Ⓒ Ⓓ Ⓔ
16. Ⓐ Ⓑ Ⓒ Ⓓ Ⓔ	32. ☐	48. ☐

NOTES

NOTES

NOTES

NOTES